建筑写意
——建筑师的创意

[英] 史蒂夫·鲍克特 著

谢 天 译

中国建筑工业出版社
CHINA ARCHITECTURE & BUILDING PRESS

著作权合同登记图字：01-2014-0566号

图书在版编目（CIP）数据

建筑写意：建筑师的创意 /（英）鲍克特著；谢天译.
—北京：中国建筑工业出版社，2015.3
ISBN 978-7-112-17504-8

Ⅰ.①建… Ⅱ.①鲍…②谢… Ⅲ.①建筑画－作品集－英国－现代 Ⅳ.①TU204

中国版本图书馆CIP数据核字（2014）第269762号

Text and Images © 2013 Steve Bowkett. Steve Bowkett has asserted his right under the Copyright, Designs and Patent Act 1988 to be identified as the Author of this work
Translation © 2014 China Architecture & Building Press

This book was designed, produced and published in 2013 by Laurence King Publishing Ltd., London.

本书由英国Laurence King 出版社授权翻译出版

责任编辑：程素荣
责任校对：李美娜　王雪竹

建筑写意
——建筑师的创意
[英]史蒂夫·鲍克特　著
谢　天　译

*

中国建筑工业出版社出版、发行（北京西郊百万庄）
各地新华书店、建筑书店经销
北京锋尚制版有限公司制版
北京中科印刷有限公司印刷

*

开本：880×1230毫米　横1/16　印张：10　字数：256千字
2015年2月第一版　2015年2月第一次印刷
定价：35.00元
ISBN 978-7-112-17504-8
　　　　（26269）

版权所有　翻印必究
如有印装质量问题，可寄本社退换
（邮政编码100037）

本书适合于各年龄层次的建筑师

ARCHI— DOODLE

AN ARCHITECT'S ACTIVITY BOOK

谨将此书献给亨利，他不仅是一位睿智的男人、智慧的父亲，也是一位出色的木匠。

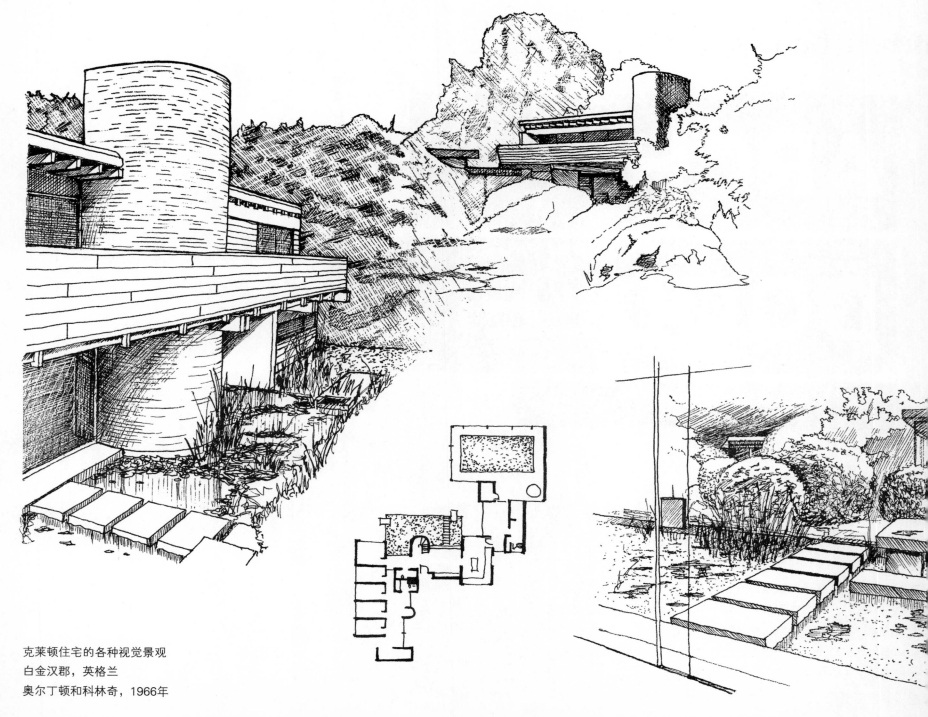

克莱顿住宅的各种视觉景观
白金汉郡,英格兰
奥尔丁顿和科林奇,1966年

troduction（导言）

本书适合于所有对建筑感兴趣的读者，尤其适合于那些喜欢绘画、涂鸦和人类建成环境的读者。本书的篇章结构是围绕一系列设计绘图练习展开，从有趣的教学练习和丰富的知识供给逐步深入到激发读者的灵感。所选取建筑和景观不仅发挥着旨在创造崭新的建筑设计形式的催化剂的作用，而且也阐述了某些奠定"现代"建筑的理念。

本书的绘图都是使用针管笔绘制（目的是为了图片清晰）的黑白插图，比较清晰简洁。鼓励读者在绘制过程中使用各种常用工具：铅笔、钢笔、颜料、蜡笔、拼贴画等等。我建议读者使用本书中的插图，在页边处甚至可以在插图上尝试着色，随意勾画，当然也可以在更大的空间里进行练习。

本书的宗旨并不在于给读者传授如何绘制建筑——市面上有很多这类书籍供其选择——本书的目的在于教读者绘制什么。因此，本书应该被视作适合放在公文包内的一本杂志、速写本或者是启蒙读本。因此，本书的前几页介绍了基本的绘图工具和技术，以供读者入门学习之用。

我衷心希望本书能给建筑师、学生、老师、家长以及他们的孩子带来帮助和灵感——但最重要的是——本书能给读者带来乐趣。

作者简介：

史蒂夫·鲍克特酷爱设计，他已在多所大学或学院教学和从事建筑实践长达25年之久，目前是英国伦敦南岸大学（London South Bank University）的高级讲师。他曾在英国皇家艺术学院（the Royal College of Art in London）和伦敦中央理工学院（the Polytechnic of Central London）接受建筑教育。

史蒂夫和他的夫人简（Jane）育有3个女儿——佐薇（Zoe）、萨迪（Sadie）和菲比（Phoebe），目前举家居住在白金汉郡，过着幸福的生活。

Equipment（工具）

本书中需要用到的基本工具

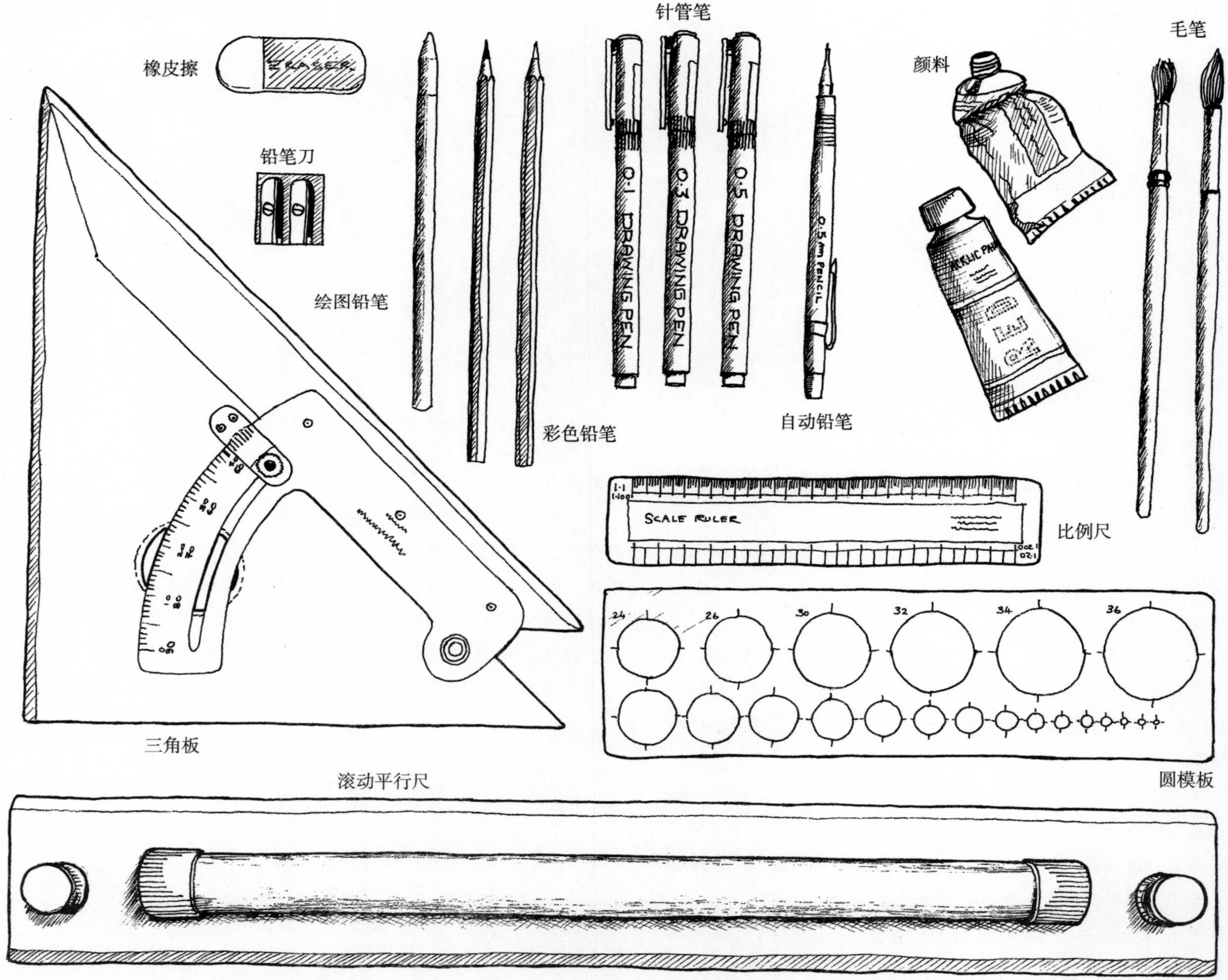

橡皮擦

针管笔

毛笔

铅笔刀

颜料

绘图铅笔

彩色铅笔

自动铅笔

三角板

比例尺

圆模板

滚动平行尺

Techniques（技法）

所展示的技法是本书绘图中必须用到的。这些简单的技法能够帮助读者表
理和形态，添加阴影，增加密度，营造各种不同的材质效果。

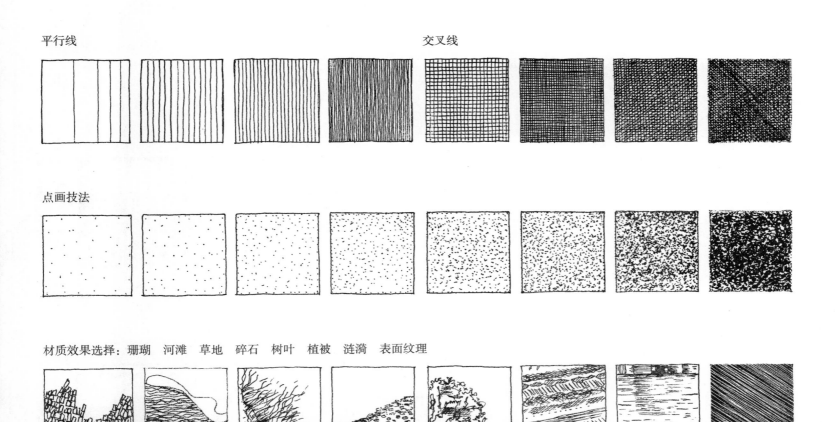

平行线　　　　　　　　　　　　　　　　交叉线

点画技法

材质效果选择：珊瑚　河滩　草地　碎石　树叶　植被　涟漪　表面纹理

Plans, Sections, Elevations, Axonometric
（平面、剖面、立面、轴测图）

这两页展示了如何绘制建筑正交投影图的方法。在最左侧的轴测图上有一根直线，正是从这里剖切可以展示轴测图下方的剖面图。剖面图下方的图是建筑的侧面或立面。对页的剖切轴测图显示了建筑顶层平面的内部房间布置。其下方的两张图分别是屋顶平面和顶层平面。

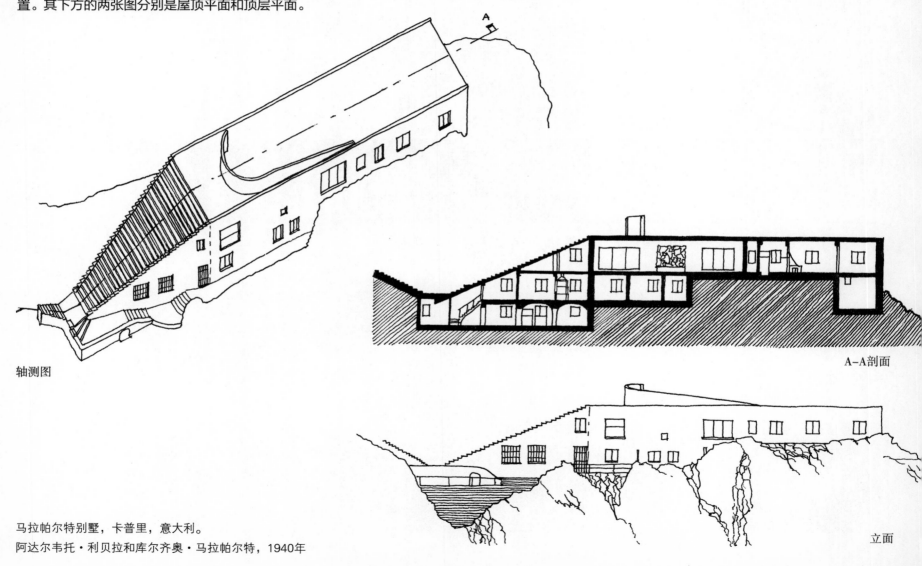

轴测图

A-A剖面

立面

马拉帕尔特别墅，卡普里，意大利。
阿达尔韦托·利贝拉和库尔齐奥·马拉帕尔特，1940年

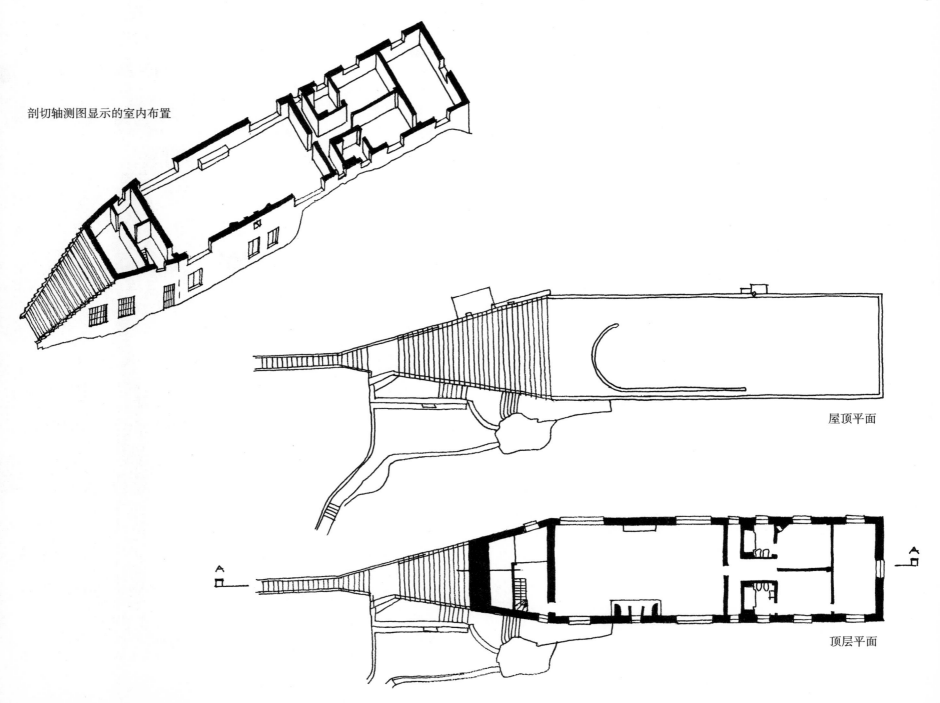

剖切轴测图显示的室内布置

屋顶平面

顶层平面

These are currently the world's **tallest buildings.**（目前世界最高的建筑）

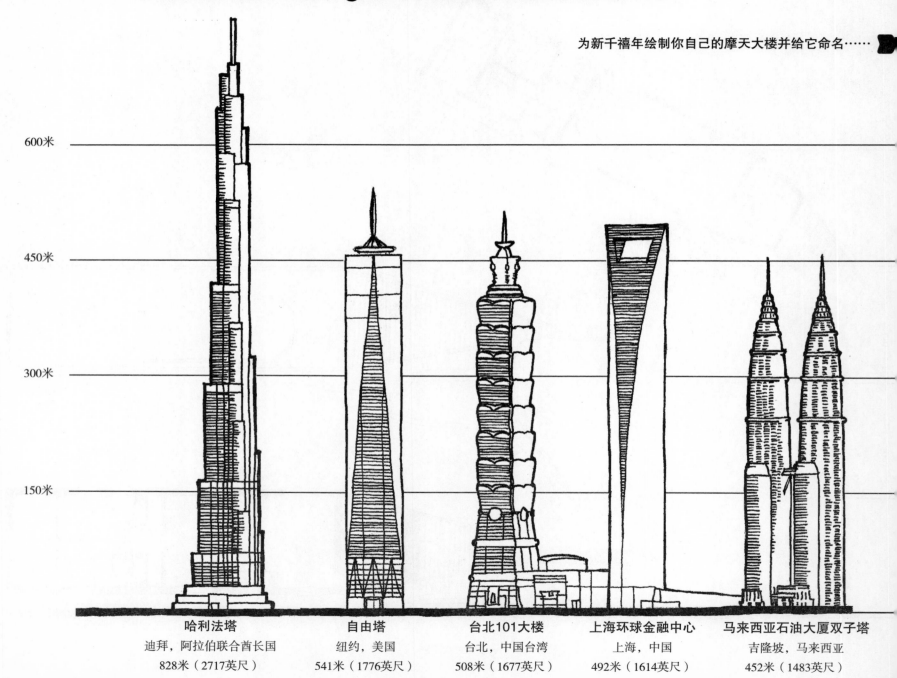

为新千禧年绘制你自己的摩天大楼并给它命名……

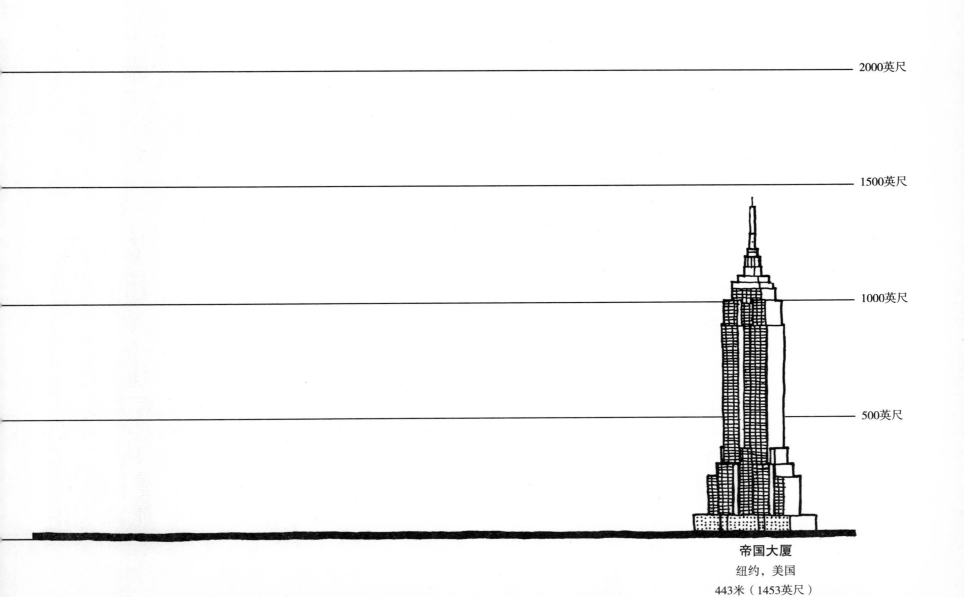

The Sony Tower has a **famous roof,** a 'Chippendale Tallboy' classical pediment top. This iconic skyscraper needs a new image and a new top. Look at other design objects such as furniture to influence the form.

索尼大厦顶部著名的屋顶—"齐本德尔式的高脚橱柜"经典的三角形山庄。这座标志性的摩天大楼需要崭新的形象和崭新的屋顶形式。请从其他家具设计中寻找新的造型灵感。

补充其他一些标志性屋顶的摩天大楼造型……

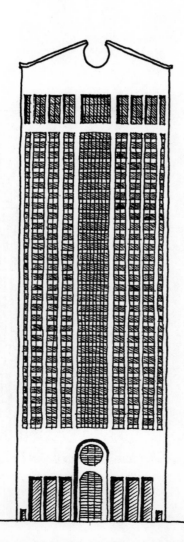

电话与电报公司大厦，纽约，美国。菲利普·约翰逊和约翰·伯奇，1984年

阿尔哈姆拉塔，科威特，SOM建筑设计事务所，2011年，
克莱斯勒大厦，纽约，威廉·凡·艾伦，1930年，
帝国大厦，纽约，施里夫，兰姆与哈蒙事务所，1931年，
碎片大厦，伦敦，伦佐·皮亚诺，2012年

14

Complete the drawings of the following **famous buildings.**

补充绘制以下著名建筑

泰姬玛哈，阿格拉，印度，
乌斯塔德·艾哈迈德·拉合里，1653年

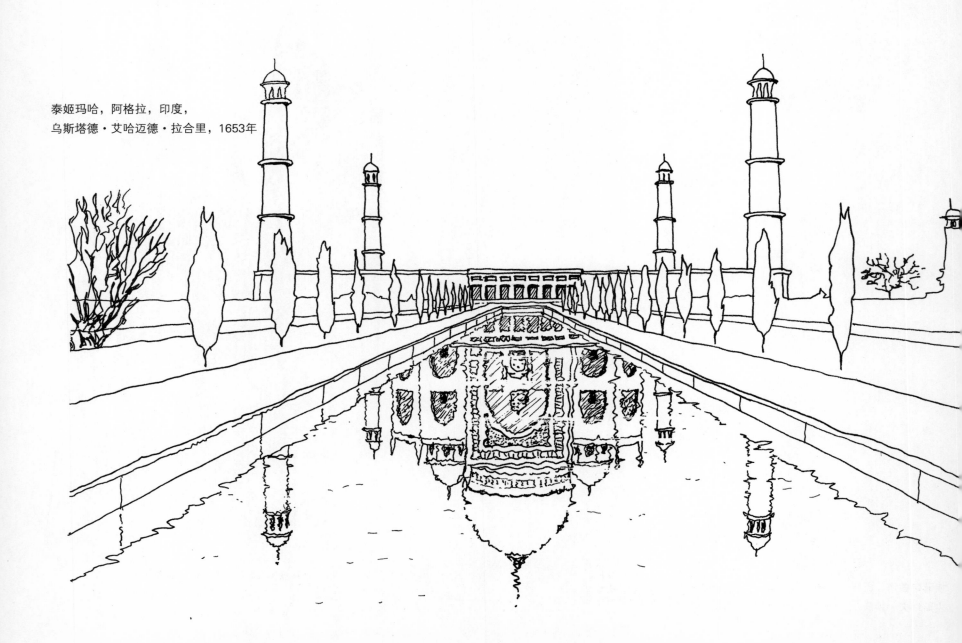

埃菲尔铁塔,巴黎,法国
古斯塔夫·埃菲尔,1889年

Cities beneath（水下城市）

海洋也许能提供未来居住的新方式，以下构造源自珊瑚造型的启示。

在书中空白处设计你的水下世界……

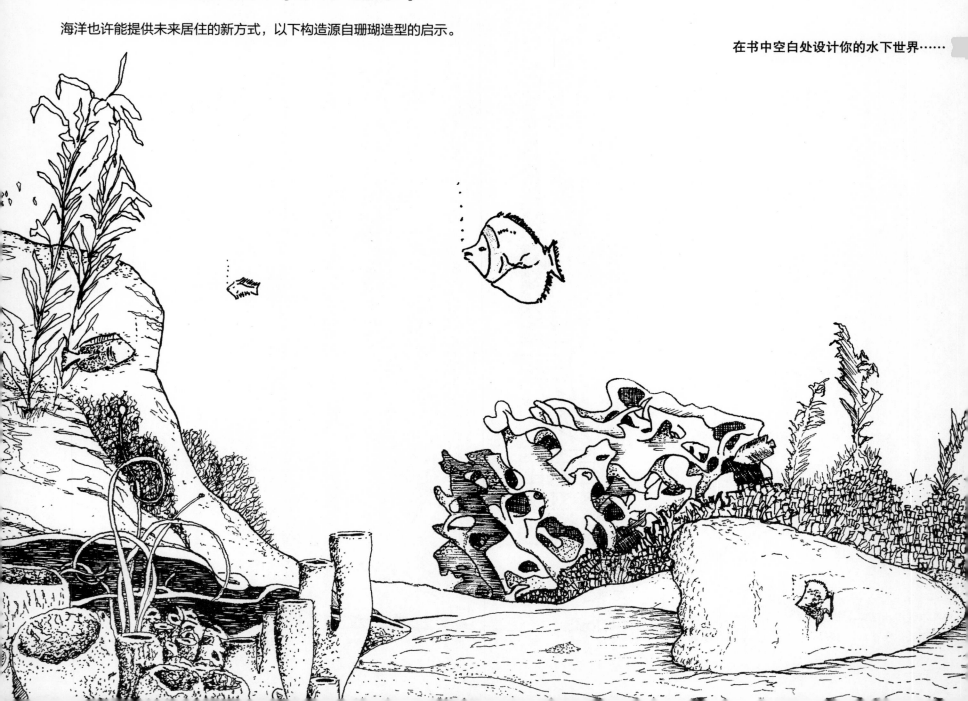

18

Here are a variety of different **bridge designs.**

各种不同的桥梁设计

采用这些结构形式设计一座新
桥使得城镇可以跨越海面……

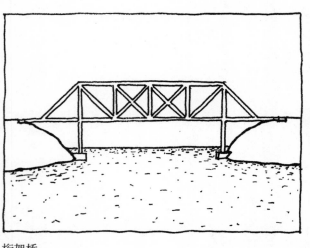

桁架桥

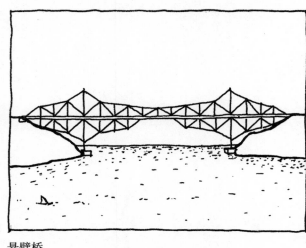

悬臂桥

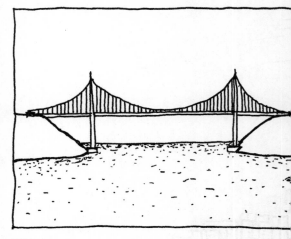

吊桥

斜拉桥

拱桥

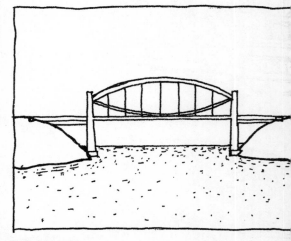

鱼腹式桁架桥

设计一座跨河大桥

十一座横跨高速公路的桥梁

In Dubai they have created a series of **new islands** along the coastline.
迪拜沿海岸线设计了许多人工岛

设计你的新沙漠天地……▶

棕榈岛，人工岛，迪拜，阿拉伯联合酋长国，
由谢赫·穆罕默德·本·拉希德·阿勒马克图
姆构思，棕榈岛集团开发，2003年8月

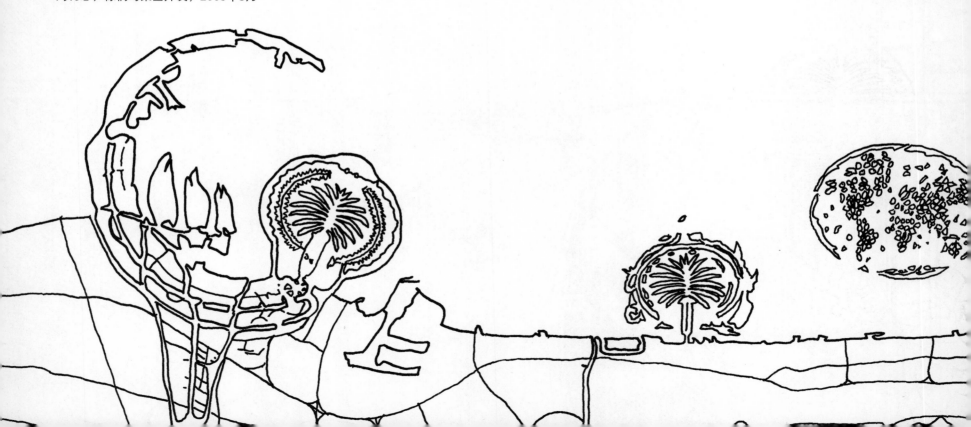

'The architect of the future will **build imitating Nature'**
— Antonio Gaudí

未来建筑师将建造自然主义建筑
——安东尼奥·高迪

设计模仿自然界生物的建筑外立面……

巴特罗公寓，巴塞罗那，西班牙
安东尼奥·高迪/何塞普·玛丽亚·茹若尔，1906年

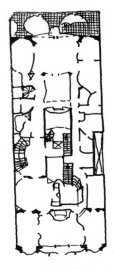

平面图

Here are six examples of how the same room can be altered to **manipulate light** and space using different forms of window opening.

本页的6个实例解释了同一房间通过不同形式的开窗方式的变化从而调控光线和空间

采用相同室内空间比例绘制你自己的构思……

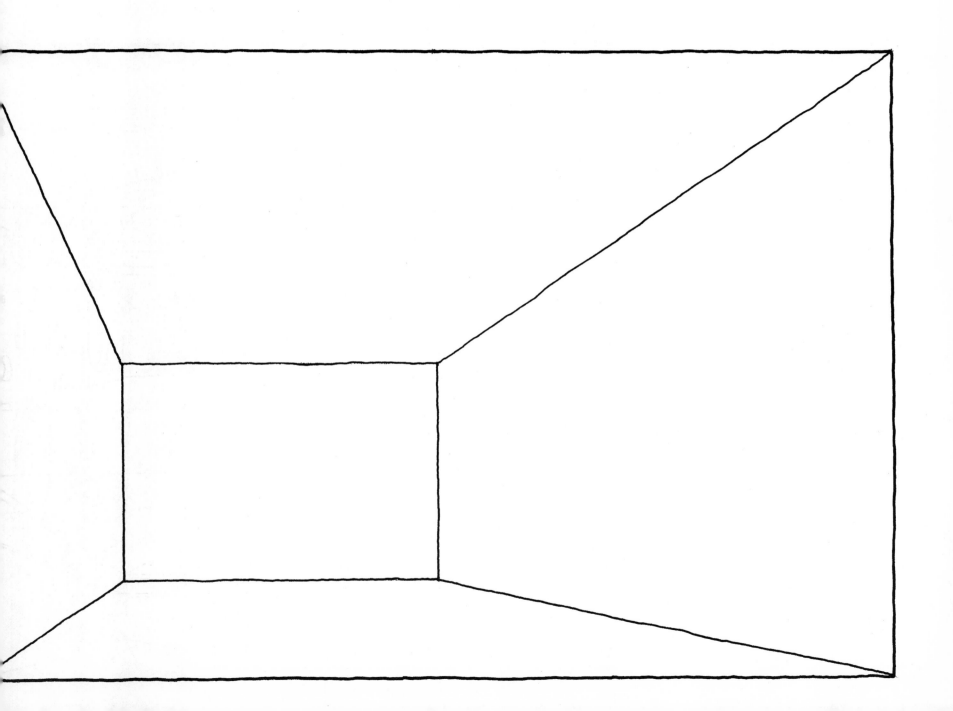

Theis and Khan designed a new **sacred space** while converting a London church, using the concept of a beam of light pouring through the roof.

泰斯和卡恩建筑事务所设计了一类新型的神圣空间用于改造一座伦敦的教堂，它采用了从屋顶倾泻的光束的理念。

采用现有的建筑外形，设计并绘制一种通过光线创造形式的新空间理念……

Lumin United Reform 教堂，伦敦，英格兰
泰斯和卡恩建筑事务所，2008年

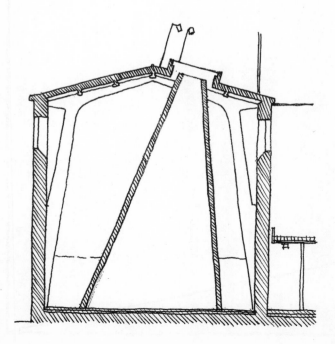

剖面展示的"光束"墙体和内部空间

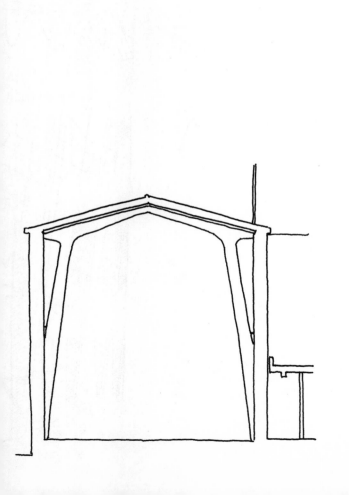

When designing the **roof of a building** here are a variety of forms to consider.

设计一栋建筑的屋顶时，可以考虑不同的造型

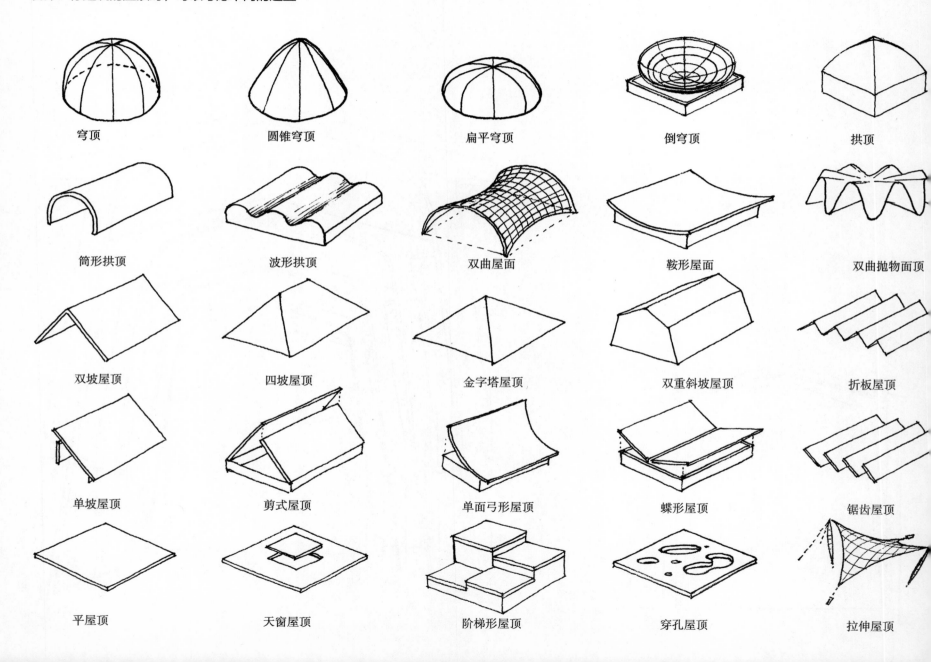

否能构思其他的屋顶形式？

The **Sydney Opera House** is dominated by its shell-like roof structure sitting on a large raised podium. Design your own roof, considering how the interior acoustics of the building might influence the external roof shape.

悉尼歌剧院因其壳状屋顶形式而出名，它坐落于一块巨大的抬高基座之上。设计你自己的屋顶，研究建筑的室内声学需求如何影响外部屋顶的形式。

设计一种可以由基座支撑的屋顶形式……

悉尼歌剧院，悉尼，澳大利亚，约恩·伍重，1973年

Characterized by their **symbolic roof forms,** representing equilibrium and balance, the civic buildings for Brazil's new capital Brasília were designed by Oscar Niemeyer.

巴西新首都巴西利亚的议会大厦由奥斯卡·尼迈耶设计,它具有标志性的象征性屋顶形式,代表了均衡与平衡。

提出其他屋顶造型的建议,它可以象征21世纪巴西利亚的进步……

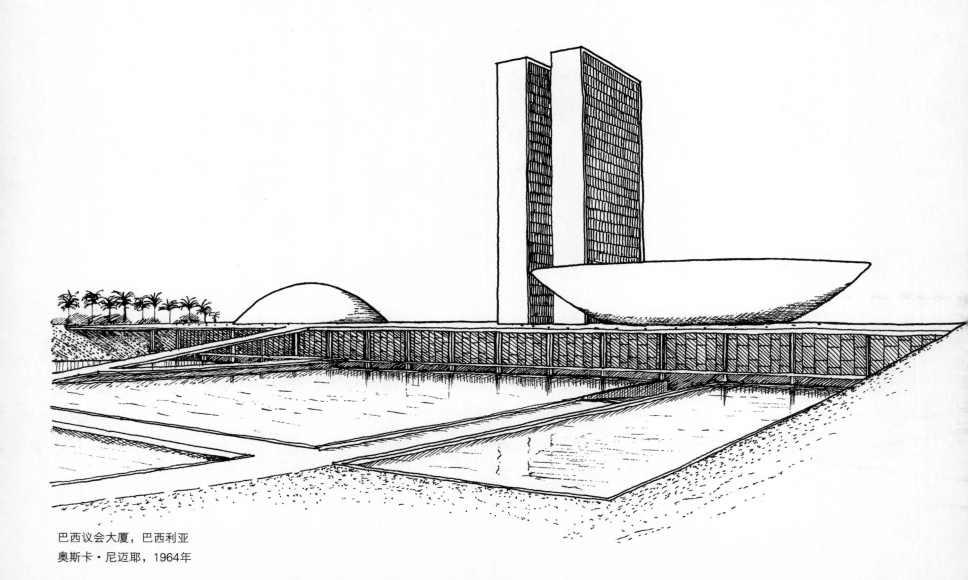

巴西议会大厦,巴西利亚
奥斯卡·尼迈耶,1964年

Austrian architectural group **Coop Himmelb(l)au** often begin projects with an intuitive sketch that they describe as 'draw[ing] with one's eyes closed', akin to the 'automatic drawing' practised by Dadaists and Surrealists. This rooftop extension is an example of this method.

奥地利建筑事务所蓝天组合通常以直观的草图开始设计，他们称之为"闭上一只眼睛画图"，这类似于达达主义与超现实主义的"自动性绘图"。这张屋顶展开图即是这种方法的一则实例。

设定你专属的"自动性绘图"，采用这种方法画出建筑的平面和剖面。首先你也许需要绘制一张草图……

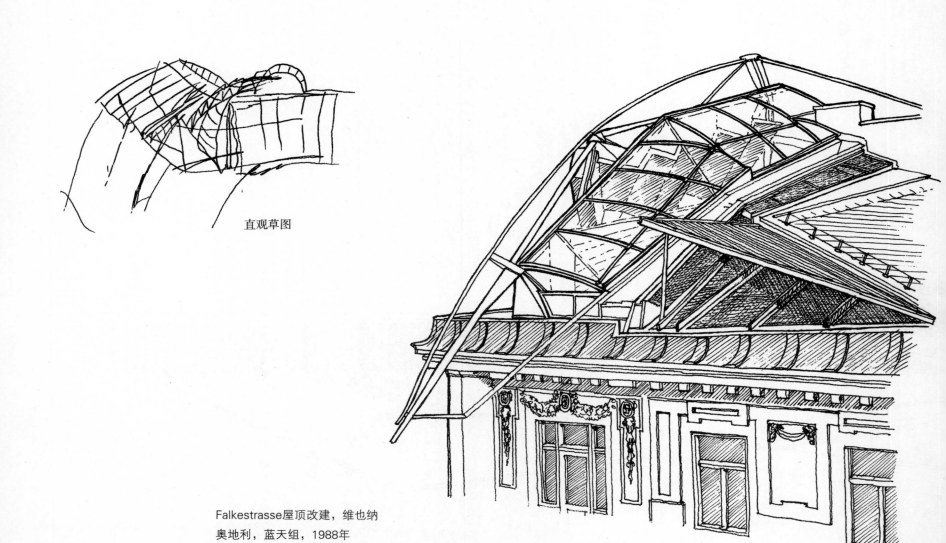

直观草图

Falkestrasse屋顶改建，维也纳
奥地利，蓝天组，1988年

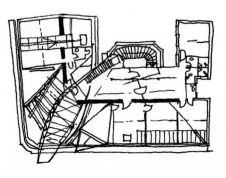

平面图

Populating planets has long been a dream of man
as he has explored the outer reaches of space.

居住于外星球一直是人类的梦想,因为人类已经探索了地球的外部空间。

在土卫六星表面设计一个失重的新居民区……

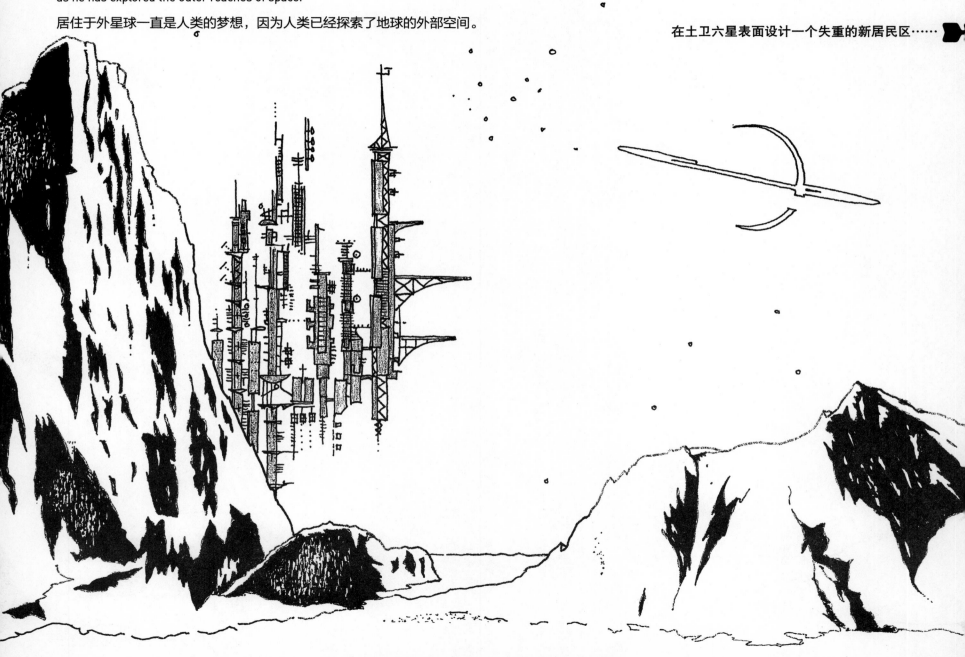

The **De Stijl** (the Style) movement in the Netherlands began in 1917, its work characterized by abstract composition and strong primary colours. The most famous exponents of De Stijl were Piet Mondrian, Gerrit Rietveld and Theo van Doesburg.

荷兰的风格派运动始于1917年，它的作品特征是采用抽象的几何构图方式以及强烈的原色。风格派最著名的代表人物有：皮尔特·蒙德里安，格里特·里特维尔德，特奥·凡·杜伊斯堡。

设计一座售卖《风格》杂志的亭子来表达这种风格的精髓……

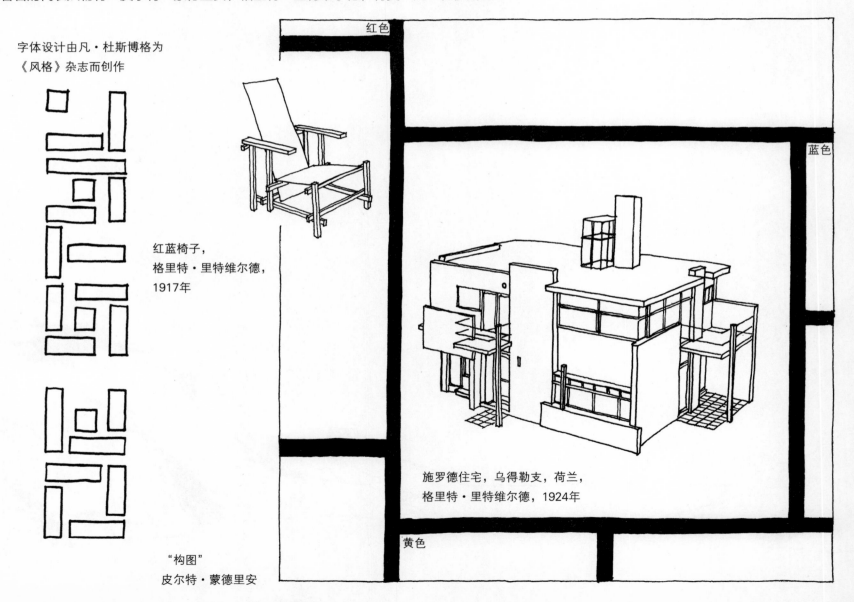

字体设计由凡·杜斯博格为《风格》杂志而创作

红蓝椅子，
格里特·里特维尔德，
1917年

红色

蓝色

施罗德住宅，乌得勒支，荷兰，
格里特·里特维尔德，1924年

"构图"
皮尔特·蒙德里安

黄色

亭子设计
拉约什·考沙克，1923年

Many architects have designed very distinctive and important **chairs** using lots of different shapes and sizes.

许多建筑师设计的椅子采用各种不同的造型和尺度，非常独特，并且具有重要意义。

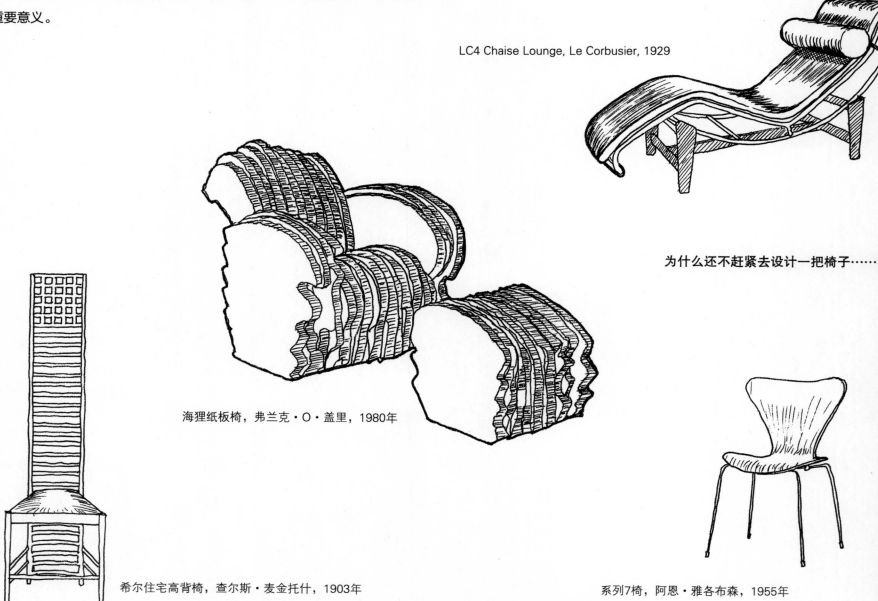

LC4 Chaise Lounge, Le Corbusier, 1929

海狸纸板椅，弗兰克·O·盖里，1980年

为什么还不赶紧去设计一把椅子……

希尔住宅高背椅，查尔斯·麦金托什，1903年

系列7椅，阿恩·雅各布森，1955年

Z字椅,格里特·里特维尔德,1934年

...you may also consider a **design for seating** for more than one person.

或许你还需要设计供多人使用的椅子

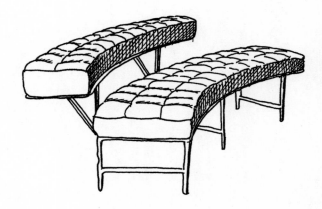

蒙特·卡罗沙发，艾琳·格雷设计，1929年

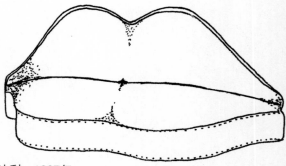

梅·韦斯特的嘴唇沙发，萨尔瓦多·达利，1937年

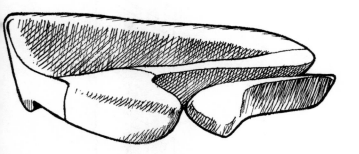

月亮系列沙发，扎哈·哈迪德，2007年

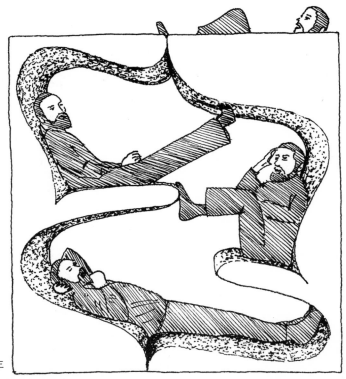

潘顿高椅，维纳·潘顿，1969年

Here are a variety of things to consider when **designing a window** and its opening.

窗户设计和开洞方式需考虑诸多要素

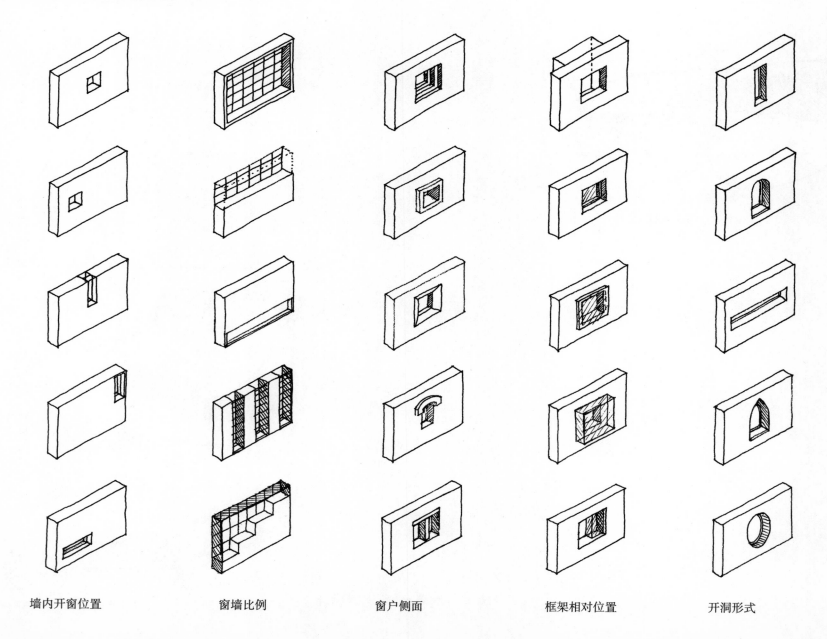

墙内开窗位置　　　窗墙比例　　　窗户侧面　　　框架相对位置　　　开洞形式

能想出更多的变化样式吗?

Each of these buildings has a different attitude to **window design:** projecting bay, window wall, free-form hole.

每座建筑都有不同的窗户设计形式：凸窗，玻璃幕墙，自由开洞。

画出其他类型的开窗方式……

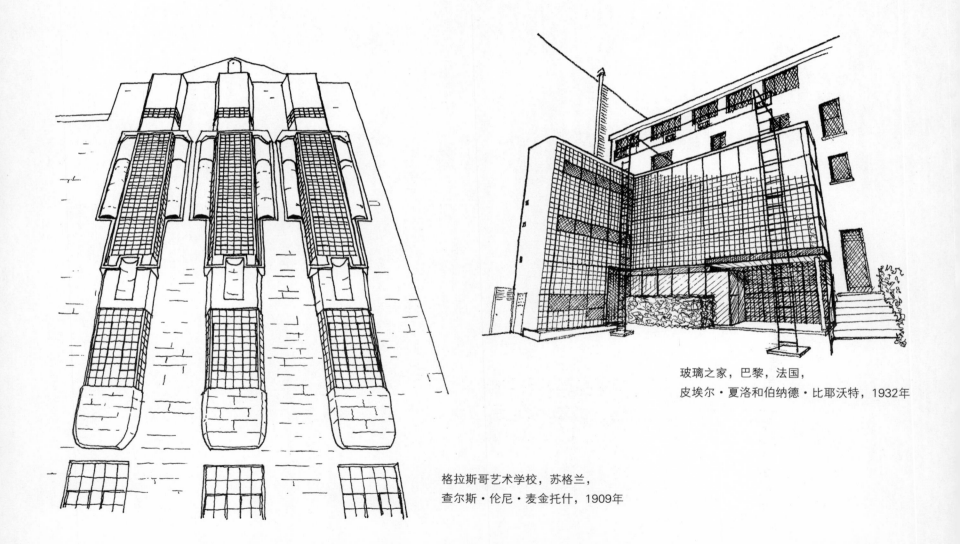

玻璃之家，巴黎，法国，
皮埃尔·夏洛和伯纳德·比耶沃特，1932年

格拉斯哥艺术学校，苏格兰，
查尔斯·伦尼·麦金托什，1909年

公共艺术建筑，
英格兰西部地区，英格兰，
威尔·艾尔索普，2008年

Different **types and shapes of window** make up the façade of this building: recessed (circular), flush (oblong), projecting (square) and clerestory (below the roofline; horizontal).

不同类型和形态的窗户构成了这座建筑的立面：凹窗（圆形），平窗（长方形），凸窗（正方形）和天窗（屋顶线下，水平的）

在同样的外墙上，用你自己的窗户和开洞方式进行设计……

维罗那人民银行，意大利，卡洛·斯卡帕，1973年

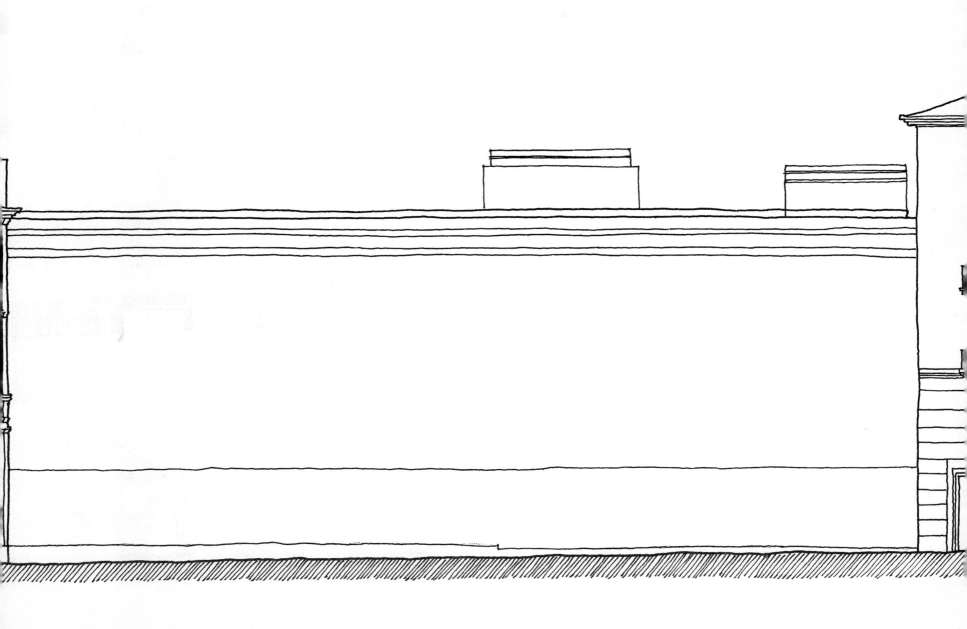

These innovative **beach huts** were created for a competition to 'Re-imagine the Beach Hut for the 21st Century'.

标新立异的海边小屋因一次竞赛诞生——"21世纪海边小屋的构想"

研究海边主题，你的竞赛设计看上去如何……

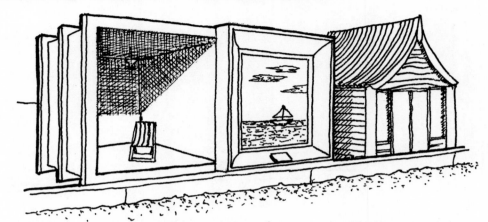

"大眼睛小屋"，费克斯和默林事务所

"外星舱体"
阿拉斯代尔·托什和加雷斯·霍斯金斯建筑事务所

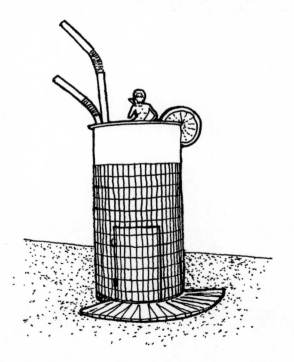

"上来看我"，迈克尔·特雷纳

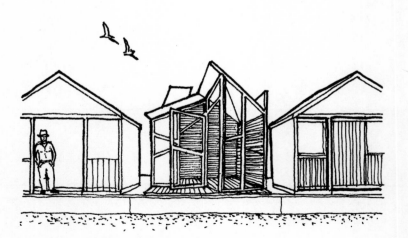

"格子间"，安德鲁·西多尔，城市设计师

"42号奶酪"
克里斯蒂安和乌尔建筑师事务所

Here is a street that is in desperate need of **'greening'.** Use some of the elements below to improve it.

这是一条非常需要"绿色"的街道,请选取以下的要素提升这条街道的品质。

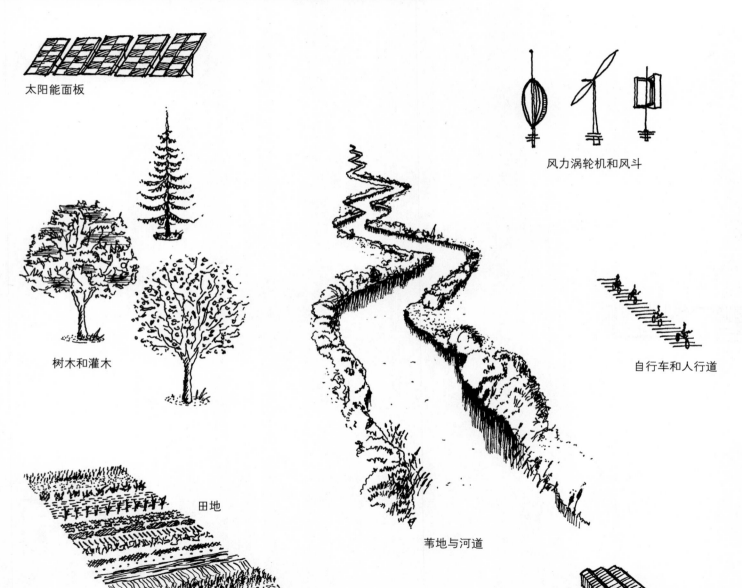

太阳能面板

风力涡轮机和风斗

树木和灌木

自行车和人行道

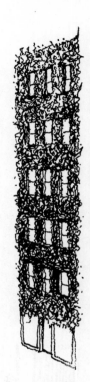

绿化墙体

田地

苇地与河道

绿色建筑

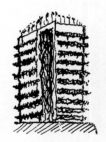

垂直栽培

These buildings have all used **recycled materials** to construct the fabric of the building.

这些建筑均使用回收材料来表现建筑的肌理

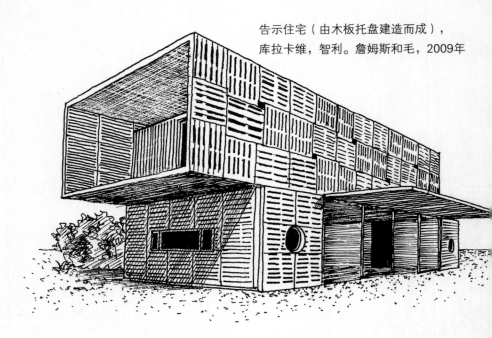

告示住宅（由木板托盘建造而成），
库拉卡维，智利。詹姆斯和毛，2009年

"亭子的短暂幸福"（33 000只啤酒箱组成），
布鲁塞尔，比利时，
SHSH建筑师事务所/V+，2008年

设计并绘制一幅用废弃材料建成的简易房屋或亭子……

露西住宅（由27000条堆叠的地毯建造而成），阿拉巴马，
美国。萨穆尔·莫克比，乡村工作室，1997年

万瓶寺（150万只啤酒瓶建造而成），
四色菊府，泰国

The **Bauhaus** school of design, founded in Germany in 1919, took a radical approach to the combined teaching of craft and design with the ambition of creating a new unity between art and technology. They explored the new techniques of industrialization through simple geometric forms.

德国包豪斯设计学校成立于1919年，它采取激进的教学方式，将工艺与设计结合在一起，目的是建立艺术和技术的整体关系。包豪斯通过简单的几何形式探索出一种全新的工业化技术道路。

用圆形、三角形和方形组合设计雕塑的形式……

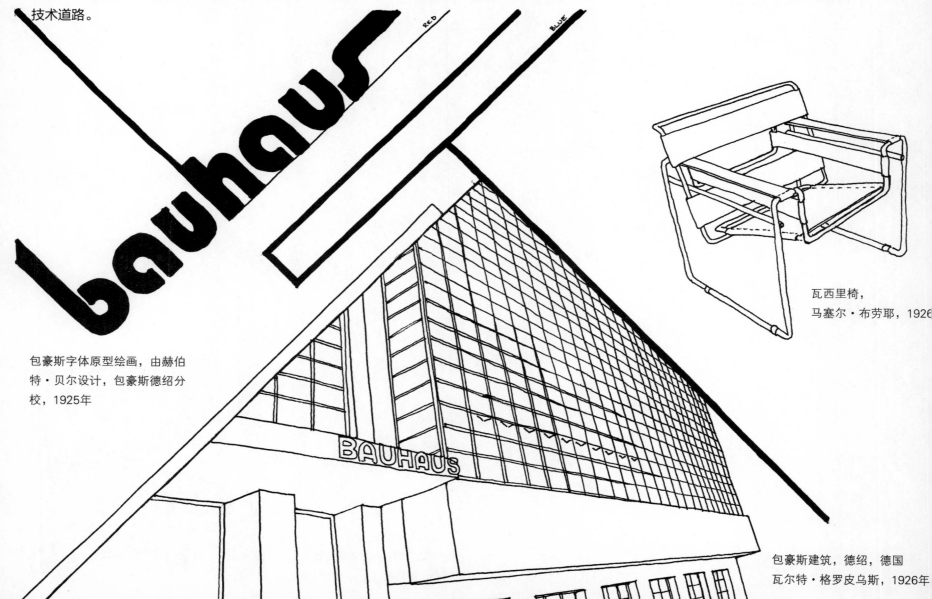

包豪斯字体原型绘画，由赫伯特·贝尔设计，包豪斯德绍分校，1925年

瓦西里椅，马塞尔·布劳耶，1926

包豪斯建筑，德绍，德国
瓦尔特·格罗皮乌斯，1926年

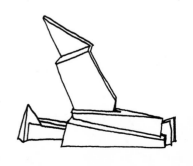

三月革命纪念碑，魏玛，德国
瓦尔特·格罗皮乌斯，1922年

Here are a various **wall structures.**
各种墙体构造

一些墙体形式，考虑它们的构造和材料……➤

Sculptural curvilinear walls of different thickness, topped off with a heavy overhanging roof, characterize the design of the Chapel of Notre Dame du Haut.

这些雕塑的弧形墙体具有不同的厚度,支撑着其上高高出挑的厚重屋顶,也成为朗香教堂的标志性建筑。

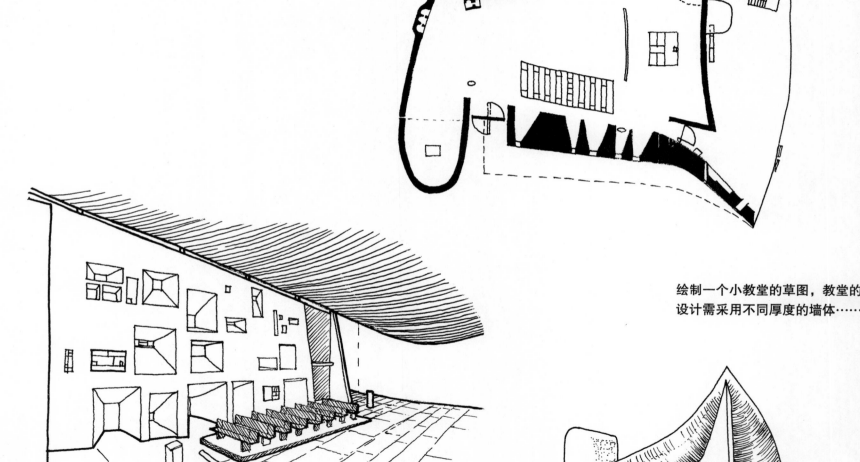

绘制一个小教堂的草图,教堂的设计需采用不同厚度的墙体……

朗香教堂,龙尚,法国
勒·柯布西耶,1954年

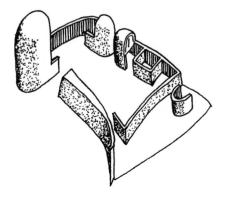

'A House is a Machine for Living in'
— Le Corbusier

住宅是居住的机器
——勒·柯布西耶

用机器美学的思想设计一座住宅,并绘制它的剖面图……

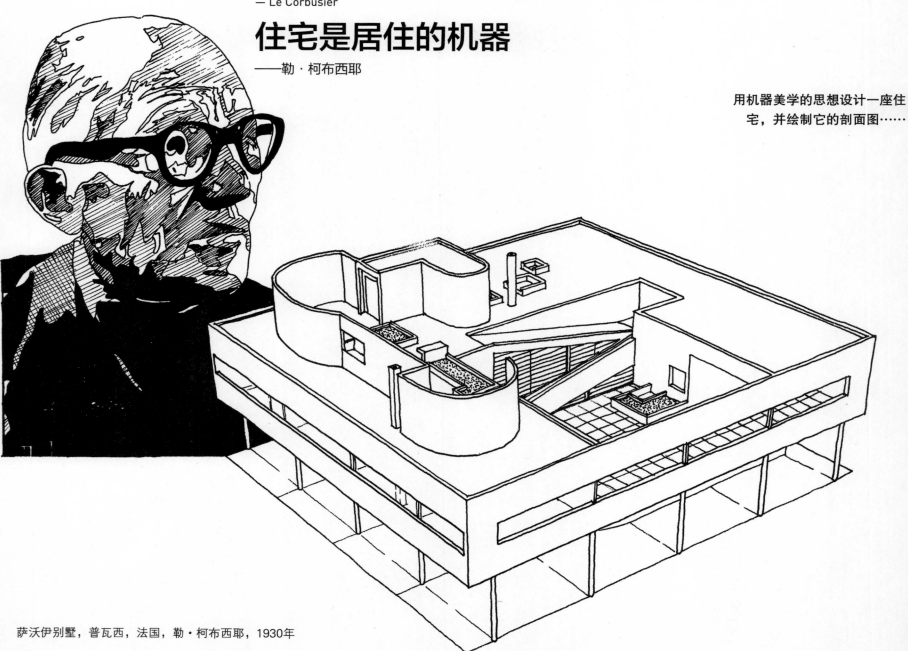

萨沃伊别墅,普瓦西,法国,勒·柯布西耶,1930年

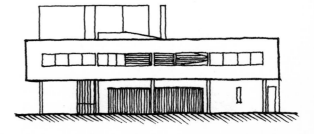

The five **columns** shown here represent the principal orders of classical architecture – Tuscan, Doric, Ionic, Corinthian and Composite.

五种柱式代表了古典建筑的主要柱式——塔司干柱式，多立克柱式，爱奥尼柱式，科林斯柱式，混合柱式

设计一款柱式……

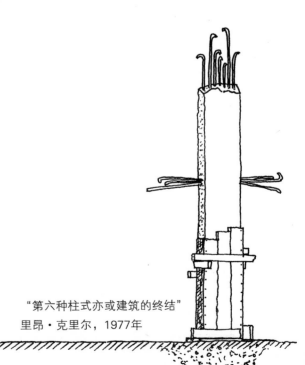

"第六种柱式亦或建筑的终结"
里昂·克里尔，1977年

Here are some **contemporary columns** to articulate how the load is transferred from the roof to the base. Try some for yourself.

当代立柱的实例清晰阐明了荷载如何从屋顶传递到基础。请设计一些立柱。

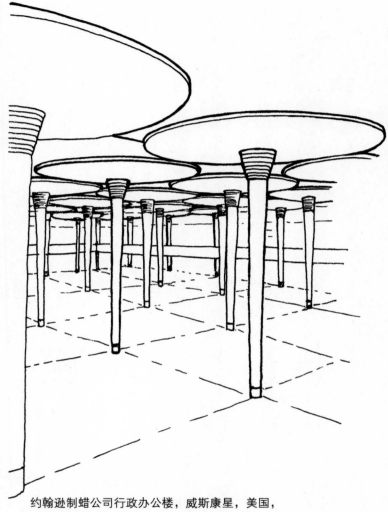

约翰逊制蜡公司行政办公楼，威斯康星，美国，
弗兰克·劳埃德·赖特，1939年

斯坦斯特德机场，伦敦，英格兰，福斯特建筑事务所，1991年

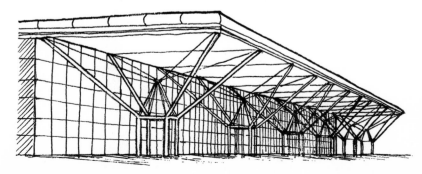

These examples show **columns and beams** expressed as one element. What ideas can you develop?

这些实例展示了柱子和梁之间的一体化关系，你能有更好的想法吗？

展览大厅，都灵，意大利，皮埃尔·路易吉·奈尔维，1949年

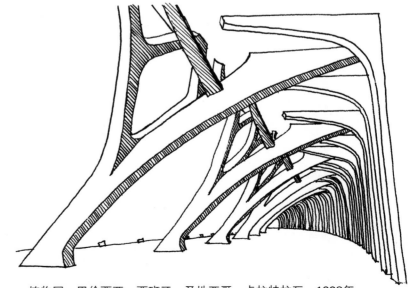

植物园，巴伦西亚，西班牙，圣地亚哥·卡拉特拉瓦，1998年

72

This copy of a comparative study of how one might resolve a **corner house** was drawn by the Luxembourgian architect Rob Krier.

这份比较分析图表明了转角建筑的多样性，它由卢森堡建筑师罗伯·克里尔绘制。

看看你能给此处的24座建筑添加多少变化……

Kinetic buildings 1:
The Sliding House by dRMM Architects has a timber skin that can be slid off to reveal a transparent layer underneath. This skin is on tracks, powered by PV solar panels, and can be moved to enjoy maximum sunlight and then retracted to retain the heat.

动态建筑1：dRMM建筑事务所设计的"滑动式建筑"外墙为木板，这层木板表皮可以移走，露出其背后镂空透明的结构。这层表皮安装在轨道上，由光伏太阳能电池板供电，它既可以移走使建筑充分享受阳光的普照，也可以移回原处防止热量散失。

设计一座简单的能与太阳互动的建筑……

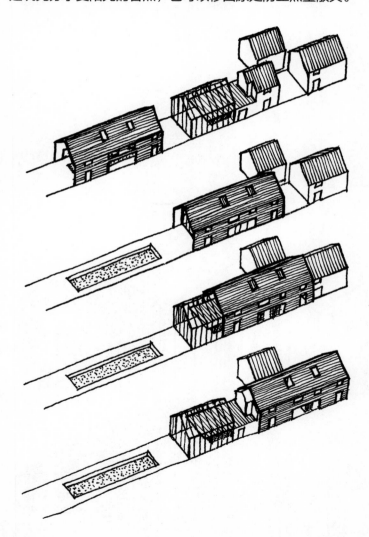

滑动式建筑，萨福克郡，英格兰，dRMM建筑事务所，2009年

Kinetic buildings 2:
Here is an example of a building that is 'in a state of permanent transformation'. Its concept is that of a simple timber box that folds open and reveals its interior. The 'GucklHupf' is used as a contemplative space as well as a place for musical performances and poetry readings.

动态建筑2：这一例子是一座"处于永久变化状态"的建筑，它的构思是一个简单的木盒子，可以折叠打开露出室内部分。"GucklHupf"既可当作沉思的空间，也可以作为音乐演出和诗歌朗读的场所。

同样依据铰链和折叠的原理设计你自己的表演空间……

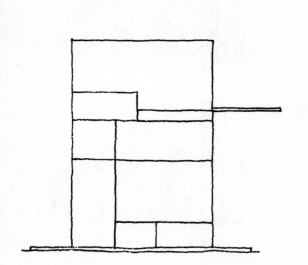
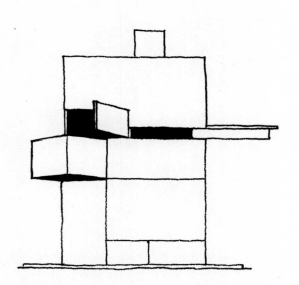
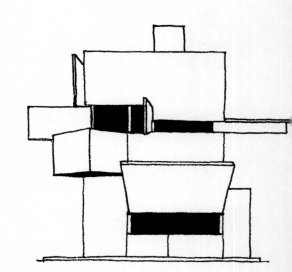

古克尔霍夫，蒙德塞，奥地利，汉斯·彼得·韦恩德勒，1993年

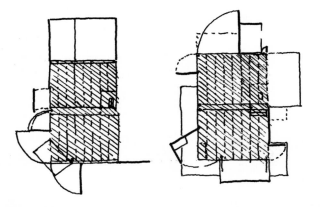

平面图

The Triangular Lodge

is a folly, and its design was a reflection of the architect's Roman Catholic faith, the number three symbolizing the Holy Trinity. This number is expressed throughout; there are three floors, three trefoil windows, three triangular gables on each side, a triangular plan, and so on.

三角屋是一种装饰性构筑物，它的设计反映了建筑师的罗马天主教信仰，数字3象征了神圣的三位一体。这一数字主导了整座建筑：3层，3个三叶草的窗户，每边3个三角形的山墙，三角形的平面等。

以某一数字，某种图案，形式或比例体系为基础，设计你自己的装饰小屋……

三角屋，北安普顿，英格兰
托马斯·特雷瑟姆爵士，1597年

平面图

The design of one of the largest parks in Paris has many **follies** laid out in a grid. The 35 follies were the architect's device for organizing and orientating visitors within the landscape. The follies were designed in the form of partly open red frameworks, waiting for both functions and events to inhabit the spaces within.

巴黎最大的公园中有许多装饰性的构筑物，它们设计为网格的形式。35座构筑物都是建筑师的一种装置产品，用以容纳和吸引景区中的游客。这些构筑物的设计形式均为局部开敞的红色框架结构体系，它们等待着功能和事件在其内部空间中诞生。

以一套框架网格为原点设计一座现代的装饰性建筑……

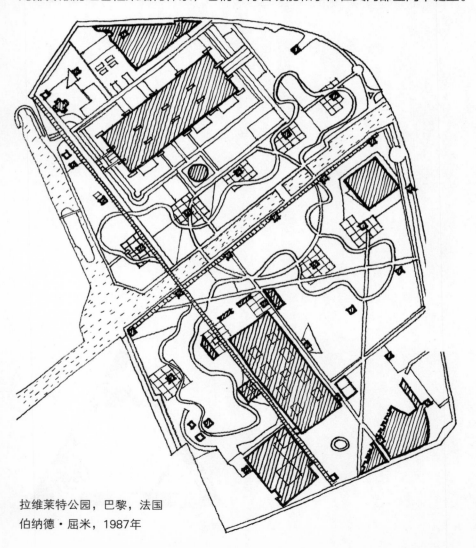

拉维莱特公园，巴黎，法国
伯纳德·屈米，1987年

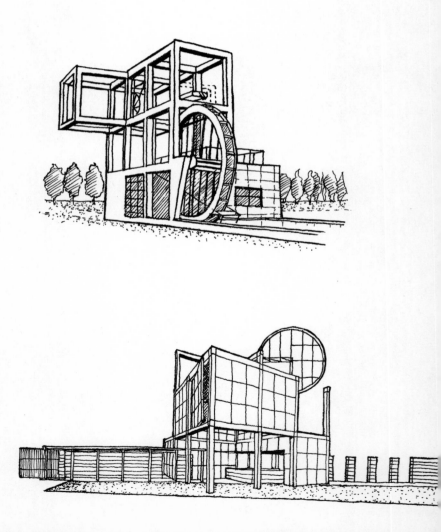

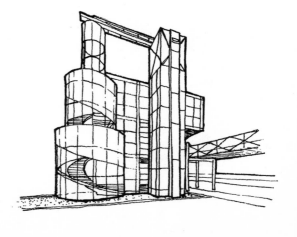

Here are some examples of unusual tree houses.（树屋实例）

补充绘制森林的空白处并在其中安置你的树屋……

Ecocoons，马蒂厄·科洛，2009年

高杉-1茶馆，日本，藤森照信，2004年

黄色树屋，奥克兰，新西兰，太平洋环境建筑师事务所，2009年

This **surreal house** in Oxford, England, has a 7.5m (25ft)-long fake shark embedded in its roof. It was erected in 1986 on the anniversary of the dropping of the atomic bomb on Nagasaki.

这座超现实建筑位于英格兰牛津，一条7.5米（25英尺）长的假鲨鱼一头扎进了建筑的屋顶。它建造于1986年长崎原子弹爆炸的周年纪念日。

绘制普通建筑的超现实画面，设定该处发生了某事件——当然也可以是你自己的房子……

"黑丁顿鲨鱼"雕塑，约翰·巴克利，1986年

飞机旅馆，哥斯达黎加，
绿色海岸，2010年

Your wealthy client wants to build a house on his **private island** but he's not sure where to build it. Think about what would be a good view, and which direction the sun and wind come from.

富有的客户想在他的私人岛屿上建一幢房屋，但他并不清楚到底建于何处。需要考虑的方面是哪一处能欣赏到美丽的风景，太阳和海风来自的方向。

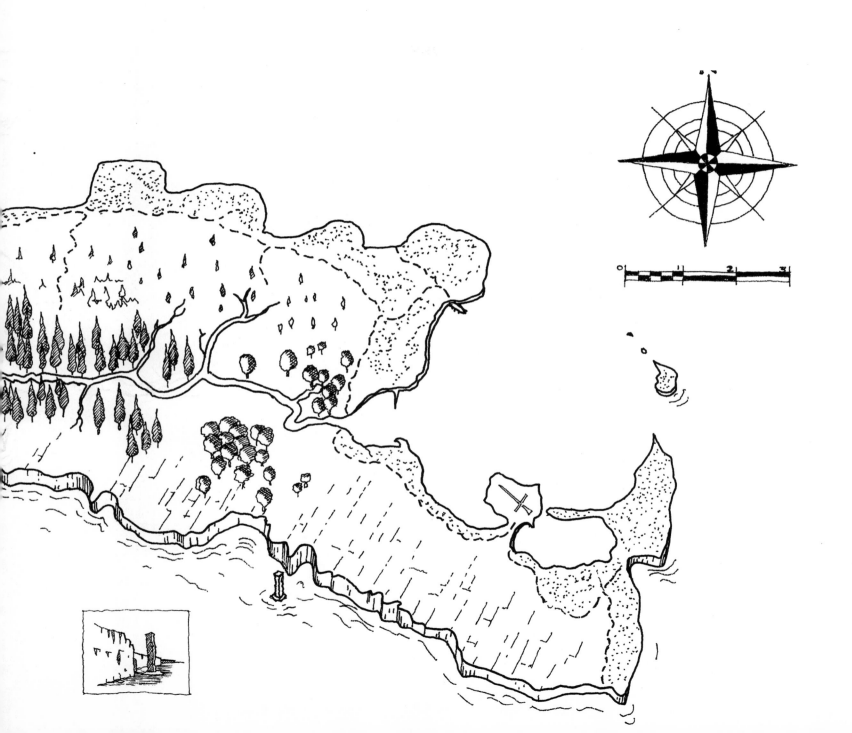

设计一座悬崖上的建筑……

一座瀑布边的建筑……

绘制一栋河边的建筑……

制山间建筑……

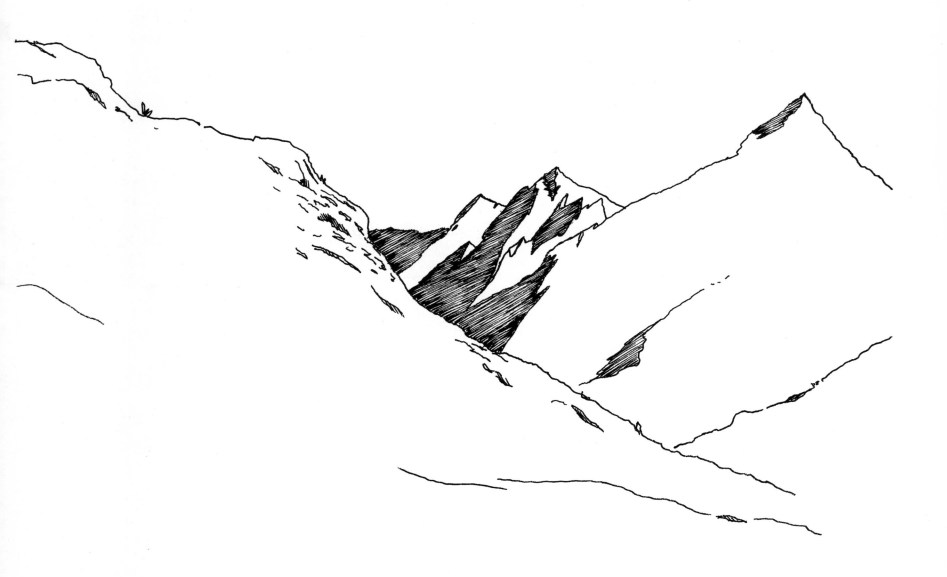

The **Constructivist** movement emerged in Russia after the 1917 Revolution. Its aim was to create an aesthetic that reflected the new industrial power of the country. Monuments like the 'Tatlin Tower' were designed using steel to celebrate 20th-century industrialization.

俄国构成主义运动发生于1917年俄国革命之后,它的目标是建立当时国家兴起的新型工业行业的审美标准。"塔特林之塔"这类纪念碑的设计使用了钢材以纪念20世纪的工业化发展。

设计一座21世纪的新纪念碑,所采用的材料和形式应反映当下的设计思潮……

至上主义构图,
卡济米尔·马列维奇,1916年

第三国际纪念碑
弗拉基米尔·塔特林,1919年

建筑机器,
雅科夫·切尔尼科夫,1931年

列宁讲坛,埃尔·利西茨基,1920年

Complete the drawings of the following **iconic buildings...**

完成以下标志性建筑的绘图

圣保罗大教堂，伦敦，英格兰，
克里斯托弗·雷恩爵士，1710年

帕提农神庙,雅典,希腊
艾克提诺斯和卡利克雷特,
公元前432年

The idea of **'servant' and 'served'** spaces in buildings was developed by the American architect Louis Kahn in the 1950s. Kahn studied Scottish castles, discovering that minor spaces – stairs and areas for cooking and washing, etc. – were contained within the walls and were like servants to the major communal 'served' areas.

20世纪50年代,美国建筑师路易斯·康提出了建筑空间中的"服侍空间"和"被服侍空间"的概念。康研究了苏格兰的城堡,发现里面存在许多小空间——楼梯间以及做饭、洗衣的空间等——它们都嵌在厚重的墙体里,如同侍者服侍主要的公共"被服侍"空间。

路易斯·康用不同色块区分四个建筑平面中的服侍空间。请区分其他建筑中的"服侍空间"和"被服侍空间"……

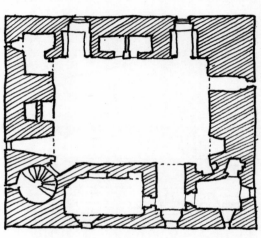
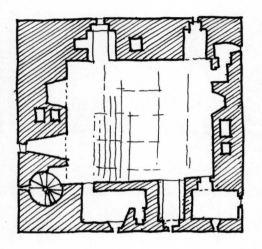
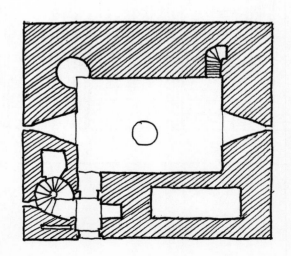
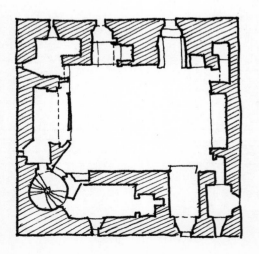

特伦多浴场，1955年

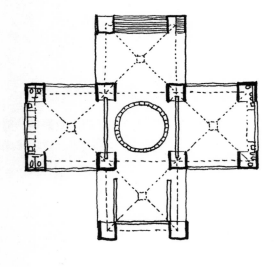

耶鲁英国艺术中心，1974年

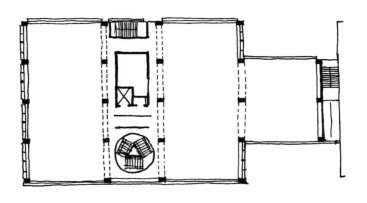

埃西里科住宅，1961年

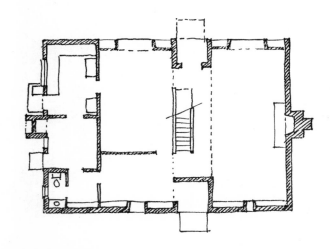

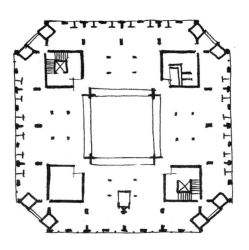

菲利普斯·埃克塞特学院图书馆

Many famous architects wear **glasses.**

许多著名建筑师都戴眼镜

设计一款新的人脸面具,剪下图样,拓上硬卡纸,然后戴上它,继续完成接下来的任务……

You have been appointed to **design a 'new look'**
for Buckingham Palace, the London residence of the Queen of England.

你的任务是为白金汉宫——英国女王的伦敦寝宫设计"新的立面"

以这张轮廓图为背景,开始你的设计创作……

白金汉宫,威斯敏斯特自治区,英格兰,
约翰·纳什,19世纪20年代

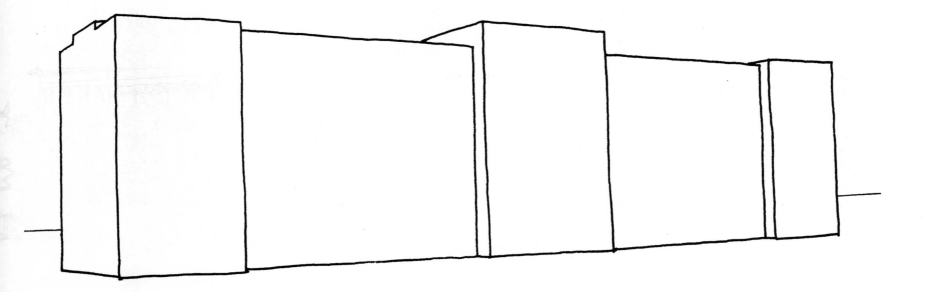

Buildings are either a **duck or decorated shed** according to architects Robert Venturi, Denise Scott Brown and Steven Izenour in their 1972 book *Learning from Las Vegas*. A 'duck' (named after a duck-shaped building that sold eggs in New York) is a building whose form tells us what its purpose is. A 'decorated shed', on the other hand, is generic in shape and can only be identified by its signage and decoration.

建筑师罗伯特·文丘里，丹尼斯·斯科特·布朗，史蒂文·艾泽努尔在其1972年出版的《向拉斯韦加斯学习》一书中，将建筑分为两种类型：鸭子或装饰的棚子。"鸭子"（以纽约一座卖蛋的鸭子造型的建筑命名）指的是建筑的形式表明了它的功能。而"装饰的棚子"指外形普通，只能通过它的招牌和装饰辨别的建筑。

绘制你自己的"鸭子"和"装饰的棚子"建筑，表明它们是通过形式还是通过招牌体现内部功能……

大鸭子，佛兰德斯，纽约，美国，1931年

金砖赌场，拉斯韦加斯，美国，1946年

"我是纪念碑"速写复印件，
文丘里和斯科特·布朗绘制，1972年

Modern **sports stadia** provide good examples of a 'Duck' or 'Decorated Shed'. However, sometimes they can be read as both. So are the stadia shown here 'ducks' or 'decorated sheds' or both?

现代体育场馆很好地阐释了"鸭子"或"装饰的棚子"的概念。但偶尔某些场馆兼具两者特征。如何区分下面的场馆是"鸭子"或是"装饰的棚子"还是兼具两者特征？

绘制这些概念所代表的体育馆草图……

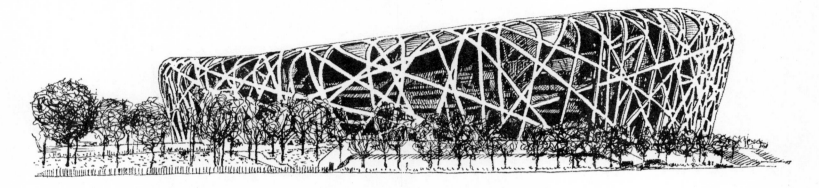

国家体育场（即著名的"鸟巢"），北京，中国
赫尔佐格和德梅隆建筑事务所，2008年

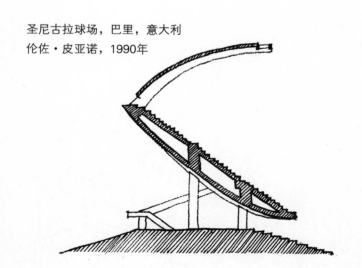

圣尼古拉球场，巴里，意大利
伦佐·皮亚诺，1990年

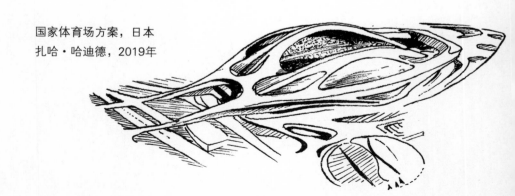

国家体育场方案，日本
扎哈·哈迪德，2019年

安联球场，慕尼黑，德国
赫尔佐格和德梅隆建筑事务所，2005年

'Less is More' "少就是多"
— Ludwig Mies van der Rohe ——密斯·凡·德·罗

设计你的小住宅……

范斯沃斯住宅,皮亚诺,伊利诺伊州
密斯·凡·德·罗,1951年

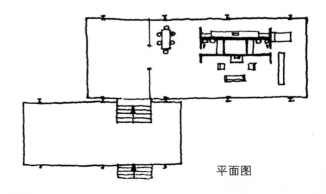

平面图

108

Here is the plan of a house for the **Berlin Building Exhibition** designed by Ludwig Mies van der Rohe and Lilly Reich in 1931.

这是柏林建筑展的展厅平面图,由密斯·凡·德·罗和利利·赖奇设计于1931年。

根据家具和设备的底图,在平面上标记你想要的烹饪、就餐、洗涤、睡眠、学习和放松区域位置……

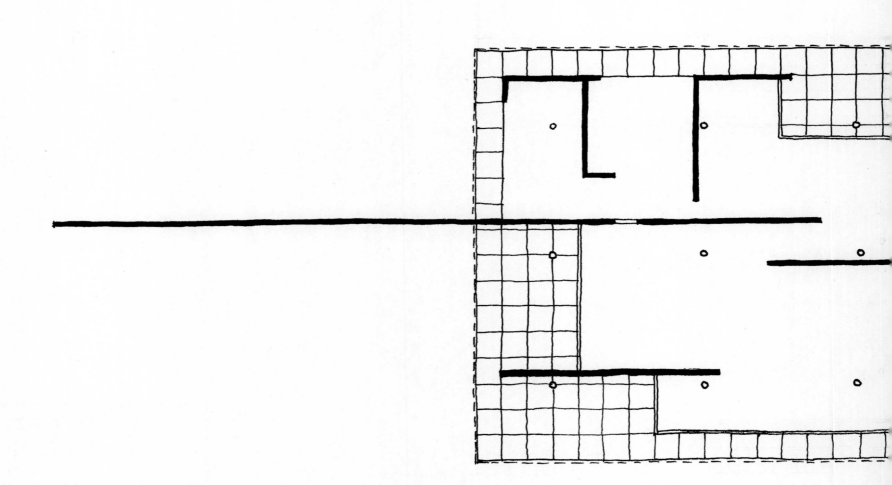

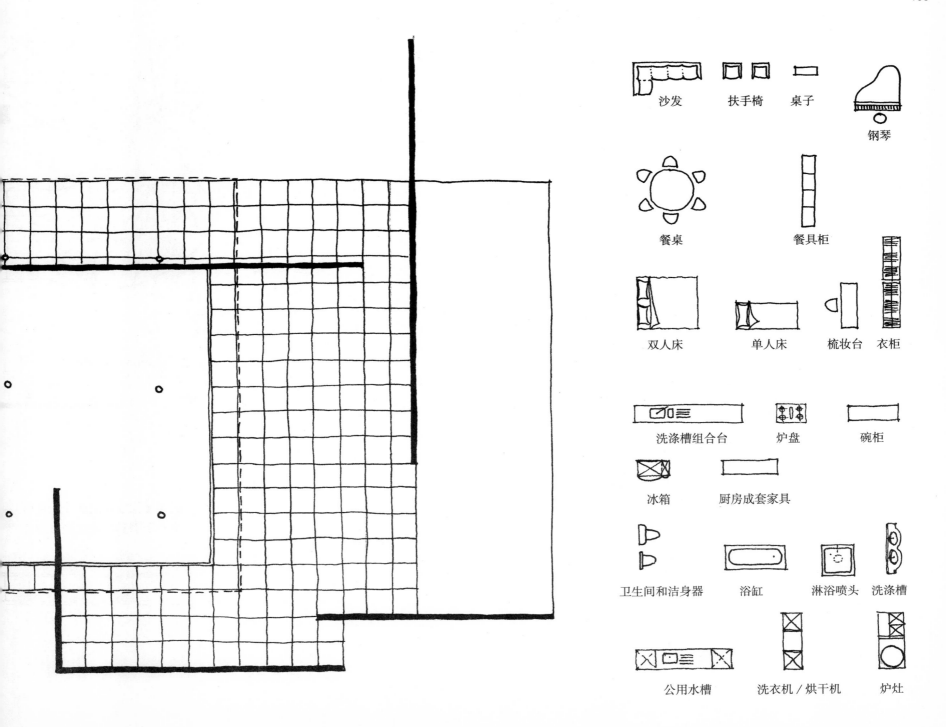

Here are some examples of **curvilinear stairs.**

曲线形楼梯实例

私人住宅，克雷乌塔，西班牙
何塞普·马里亚·尤约尔，1916年

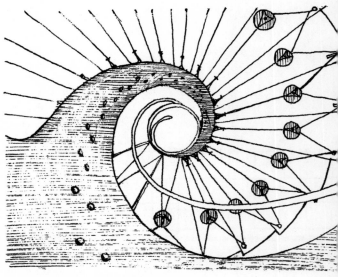

约瑟夫酒店，布拉格，捷克共和国
埃瓦·伊日奇那，2002年

绘制楼梯的草图设计，其中需表现图形和曲线的形态。

卢浮宫入口的玻璃金字塔，
巴黎，法国
贝聿铭，1989年

These **innovative stairs** were designed for a London house and incorporate a bench next to the entrance and storage leading to a bed platform.

伦敦某住宅内的创意楼梯在紧邻入口处设置了一张长凳，并在通向床铺平台的楼梯下方设置了储藏空间。

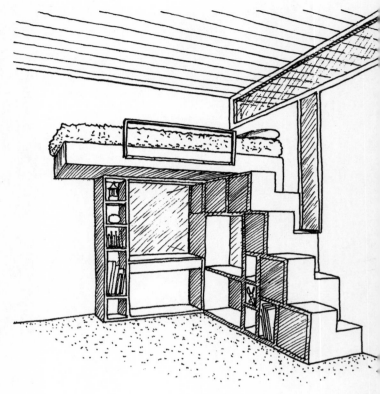

城镇住宅，伦敦，英格兰
坦卡德·鲍克特，2001年

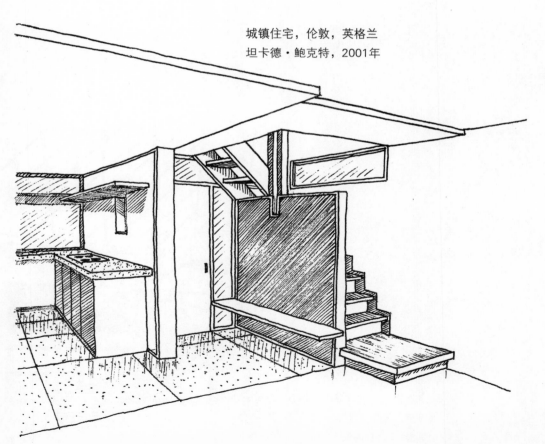

在这两处空间内绘制你自己的设计想法……

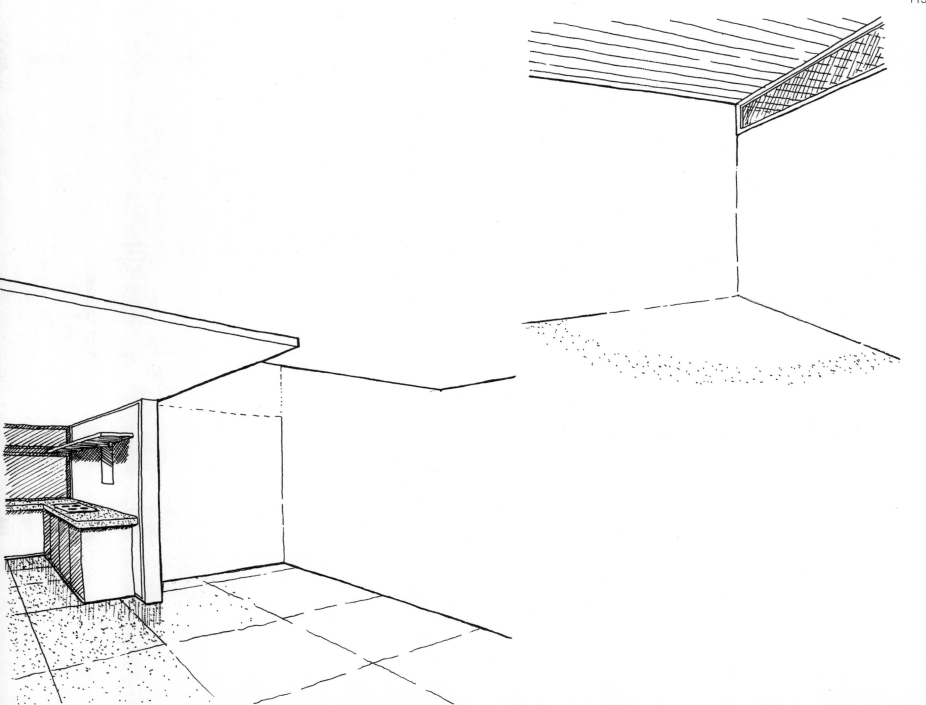

This **display staircase** appears to float within the space and provides a centre point for exhibiting artefacts in the Olivetti Showroom in Venice.

威尼斯的奥利韦蒂展厅中展示的楼梯似乎飘浮在空中,并提供了展示物品的中心区域。

你会如何设计一段通往二层夹层处的楼梯?

奥利韦蒂展厅楼梯,威尼斯,意大利,卡洛·斯卡帕,1958年

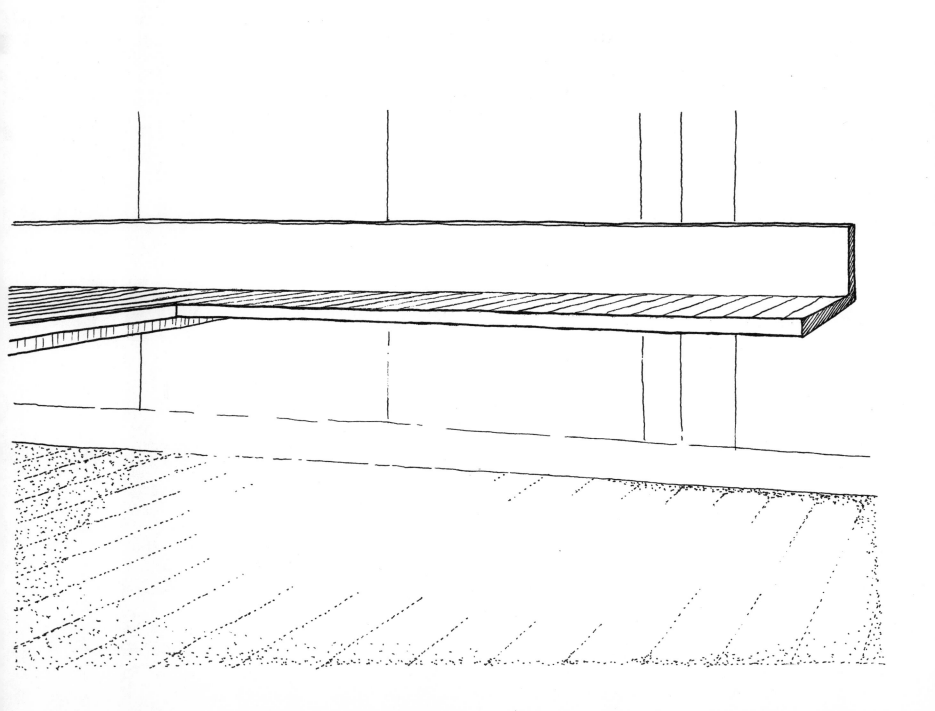

Atelier Bow-Wow Architects coined the term **'pet architecture'** for small buildings built in the left over spaces next to big buildings. If this building, known as 'the Gherkin', had a pet, would it be shaped like a pastrami sandwich? Or a mustard pot?

Bow-Wow建筑师工作室创造了"宠物建筑"一词，专用于指代那些在高大的建筑旁边剩余空间中建造的小建筑物。如果这座著名的"小黄瓜"有一个宠物建筑，它的造型是否像一块五香熏牛肉三明治？或是一个芥末罐？

选择某一著名建筑，并为它设计一个"宠物"……

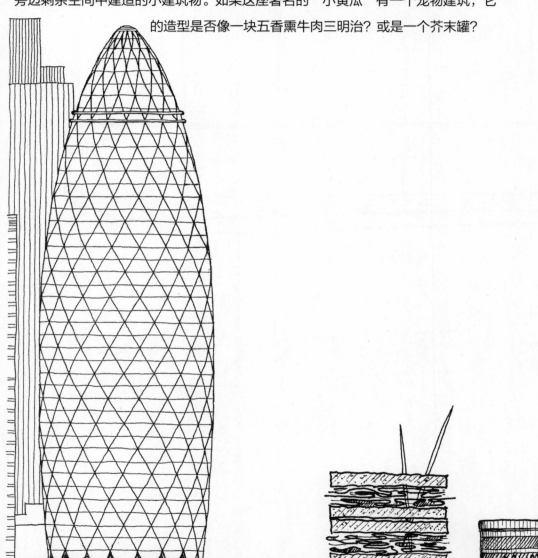

圣玛丽·艾克斯30号大楼（"小黄瓜"），伦敦，英格兰，诺曼·福斯特，2003年

In 2004, architects **Tonkin Liu** designed and built a hill-top sound sculpture overlooking the town of Burnley in Lancashire, England. Their interactive design, the 'Singing Ringing Tree', was named after the Brothers Grimm folk story.

2004年，Tonkin Liu 建筑师事务所设计并建造了一座如小山一般高的音乐雕塑，它可鸟瞰英国兰开夏的伯恩利小镇。这种互动式的设计——"歌唱之树"——以格林兄弟的民间故事命名。

设计一座与构成元素互动的雕塑造型，并给它命名……

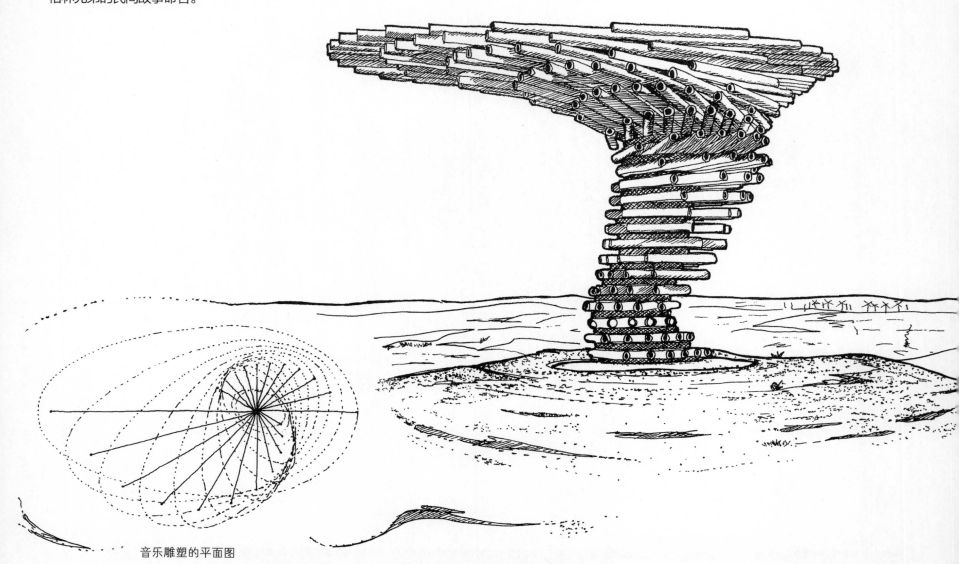

音乐雕塑的平面图

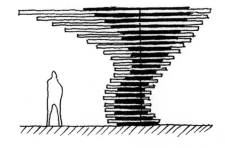

Fluid and dynamic sketching is a style developed by the German Expressionist architect Erich Mendelsohn. The sketch drawings of the Einstein Tower on the left show how the building form evolved.

流体和动态表现图是由德国表现主义流派建筑师埃里希·门德尔松创造的表现风格。左侧的爱因斯坦天文台的草图展示了建筑形式的演变过程。

用毛笔、墨水和绘图铅笔绘制一座小塔的造型演变草图……

爱因斯坦天文台,波茨坦,德国
埃里希·门德尔松,1921年

122

These are all examples of buildings with **unusual chimneys.**

奇异的建筑烟囱实例

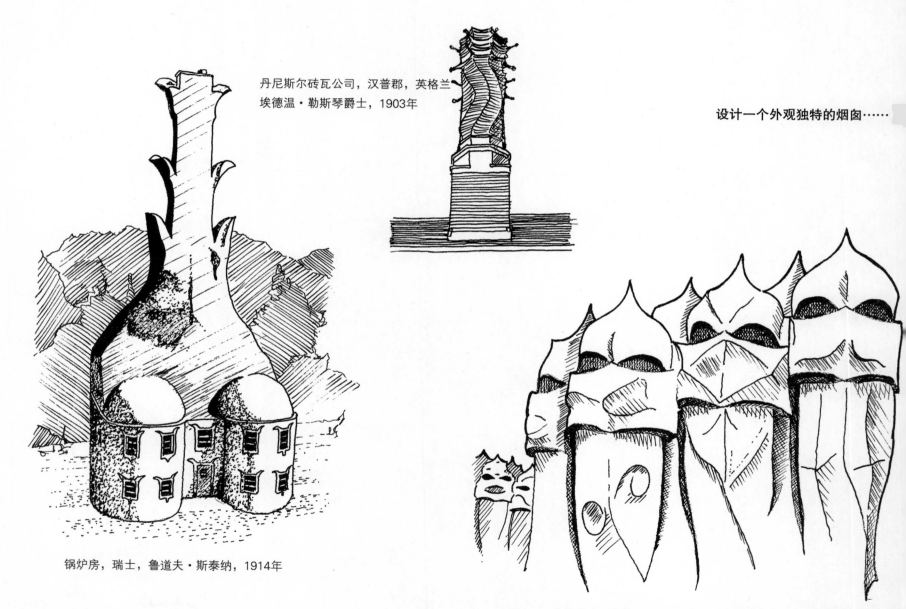

丹尼斯尔砖瓦公司，汉普郡，英格兰
埃德温·勒斯琴爵士，1903年

设计一个外观独特的烟囱……

锅炉房，瑞士，鲁道夫·斯泰纳，1914年

米拉公寓，巴塞罗那，西班牙，安东尼奥·高迪，1912年

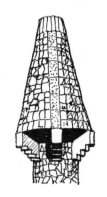

帕劳·居埃尔公园，
巴塞罗那，西班牙
安东尼奥·高迪，1890年

124

The **Borneo-Sporenburg** housing development in Amsterdam, conceived of by West 8 Architects, invited a number of architects to design terraced houses within a 5m (16ft)-wide and 12m (39ft)-long plot.

阿姆斯特丹的Borneo-Sporenburg居住区开发项目由west 8建筑师事务所设计方案，邀请了许多建筑师在5米（16英尺）宽，12米（39英尺）长的地块中设计排屋式住宅。

在空白处设计并画出面对运河的新住宅的立面……

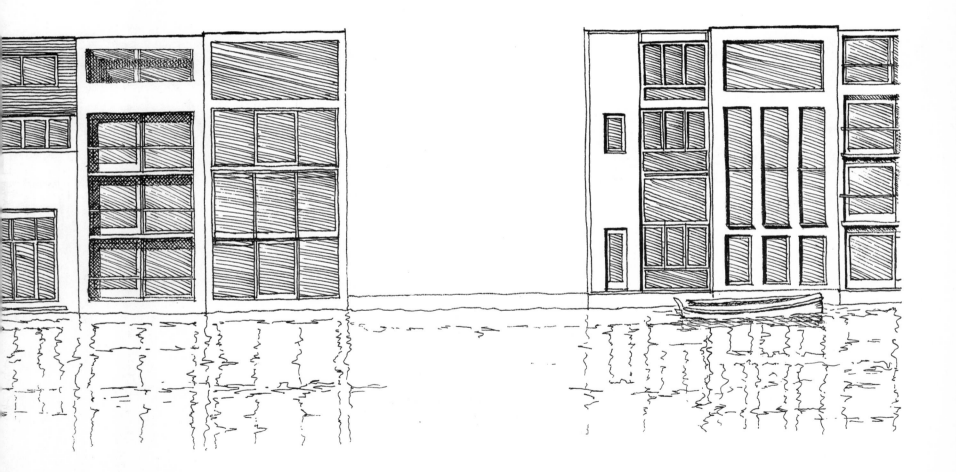

This **spiral-shaped house,** designed by the American architect Bruce Goff, is a dwelling that brings plants, pools and the surrounding countryside into the interior.

这座螺旋状的房屋由美国建筑师布鲁斯·戈夫设计，它将植物，水池和乡村景色引入到室内。

设计一座螺旋形的住宅……

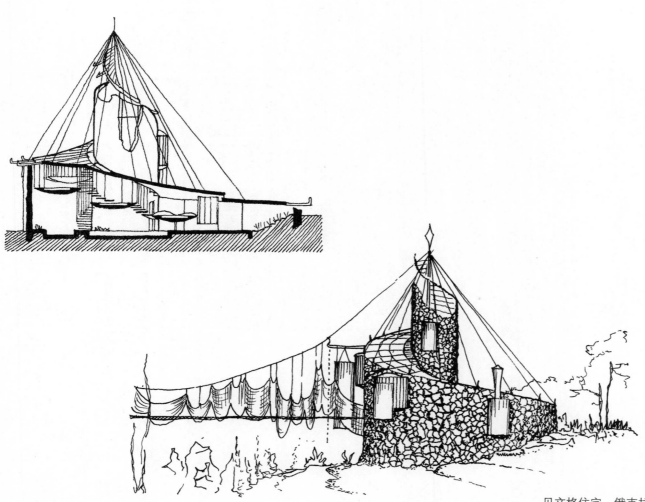
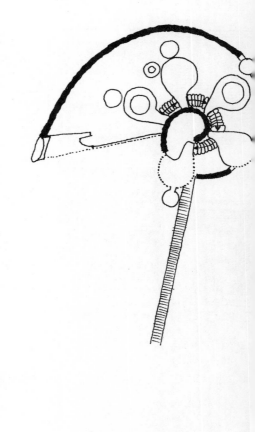

贝文格住宅，俄克拉荷马州，美国，布鲁斯·戈夫，1955年

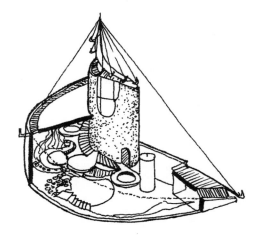

The **entrance** is a major consideration when designing a building. This threshold not only provides a way into an interior space but can also signify shelter, security, function, importance and opulence.

入口是建筑设计的一个主要考虑因素，入口不仅是进入室内的一条通道，还表明了建筑的庇护所、安全性、功能、重要性和豪华程度等特征。

贝朗榭公寓，巴黎，法国
埃克特·吉马赫，1898年

Schullin珠宝店，维也纳，
奥地利，汉斯·霍莱因，1974年

绘制并设计你附近建筑的新入口……

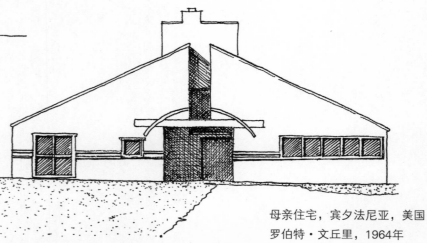

母亲住宅，宾夕法尼亚，美国
罗伯特·文丘里，1964年

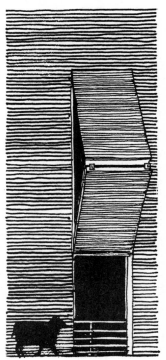

抚爱式农场，阿尔梅勒，荷兰
70F建筑事务所，2008年

When designing a **door handle,** careful consideration should be given to comfort, material, strength and style.

门把手设计需仔细考虑它的舒适性、材料、强度和风格。

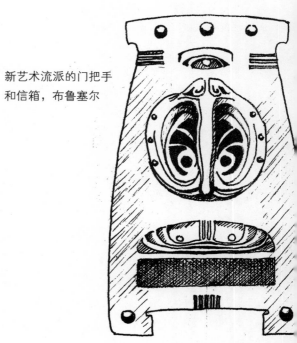

新艺术流派的门把手和信箱,布鲁塞尔

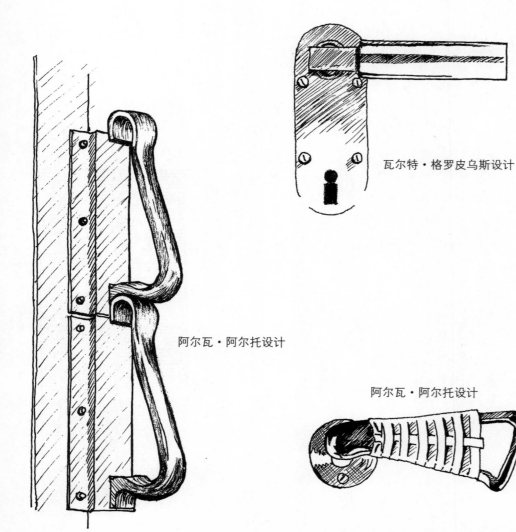

瓦尔特·格罗皮乌斯设计

阿尔瓦·阿尔托设计

阿尔瓦·阿尔托设计

绘制不同场合的各种门把手草图……

扎哈·哈迪德设计

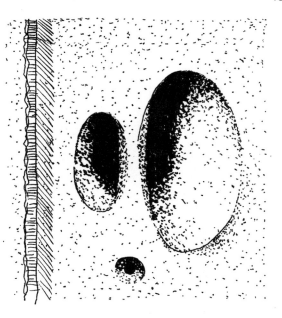

彼得·卒姆托设计

Many architects have dedicated their time to designing **buildings for animals.**

许多建筑师致力于为动物设计居所。

从动物王国中挑选一种动物，并为它设计居所……

企鹅泳池，伦敦动物园，伦敦，
贝托尔德·吕贝金的泰克松建筑小组，1934年

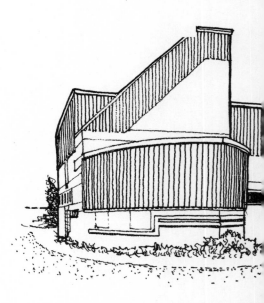

古特·加考农庄，德国，胡戈·哈林，1926

草原住宅，鹿特丹动物园，新西兰，
LAM建筑事务所，2009年

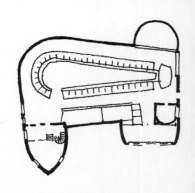

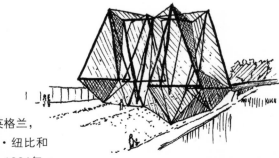

斯诺登鸟舍，伦敦动物园，英格兰，
锡德里克·普赖斯，弗兰克·纽比和
安东尼·阿姆斯特朗·琼斯，1964年

This **square house** was designed by American architect Charles Moore for himself. Made from timber, its design is based upon a series of squares within squares.

美国建筑师查尔斯·摩尔为自己设计的方形住宅，它由木材建造而成，它的设计理念是一系列方形套方形。

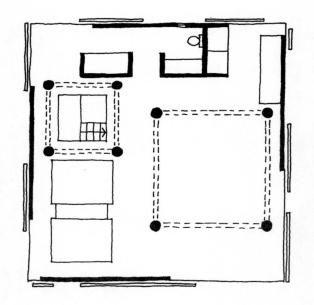

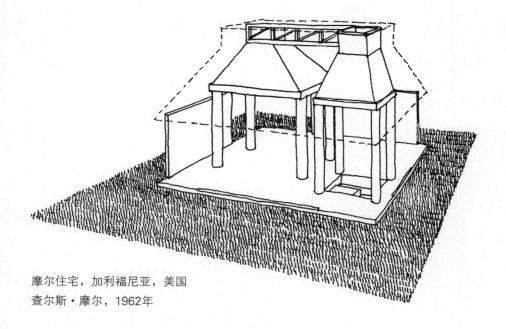

摩尔住宅，加利福尼亚，美国
查尔斯·摩尔，1962年

以方形为原型，设计一座度假小屋……

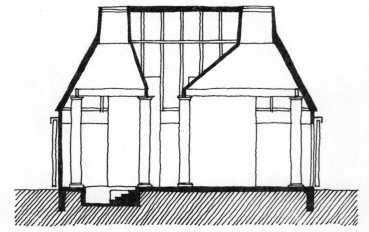

135

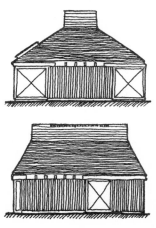

Architects have always been fascinated by exploring **rhythm and repetition** within the design of their buildings. These two examples of libraries by Finnish architect Alvar Aalto show how he changed the geometry between the fan-shaped reading areas and the rectilinear administration areas.

建筑师总是在建筑设计中着迷于韵律和重复。芬兰建筑师阿尔瓦·阿尔托的两个图书馆实例展示了他如何在扇形的阅读区域和垂直式行政区域之间进行几何图形转换。

绘制小型公共图书馆的草图，表示你将如何应用这一原理……

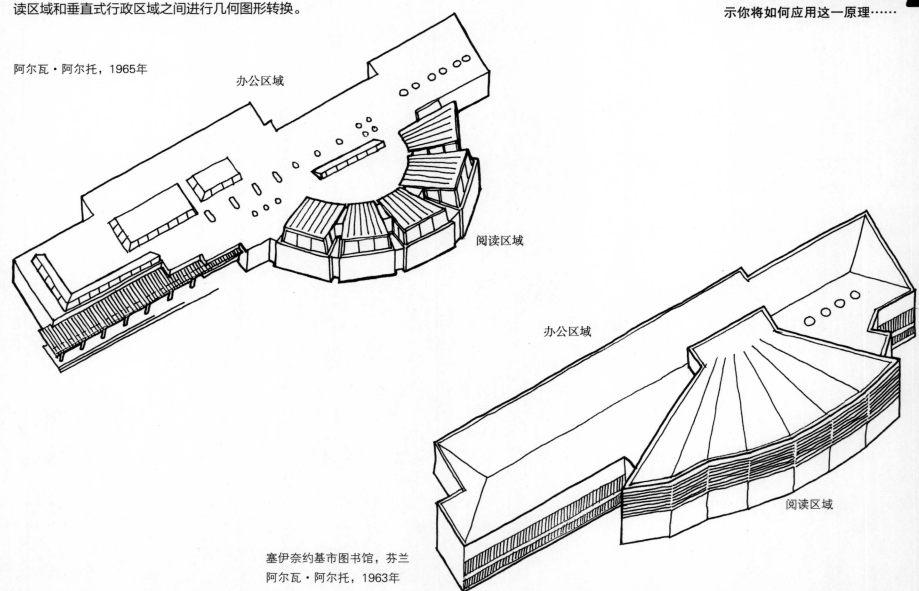

阿尔瓦·阿尔托，1965年

办公区域

阅读区域

塞伊奈约基市图书馆，芬兰
阿尔瓦·阿尔托，1963年

办公区域

阅读区域

图示韵律和重复的原理

Imagine if the owners of the **Pompidou Centre** in Paris decided that many of the building's famous exposed services were not required anymore. What should they put in their place?

设想一下，如果巴黎蓬皮杜艺术中心的业主决定不再需要那些招摇的外露设备，那他们应该在那里摆放什么呢？

在提供的空白处画出你的构思……

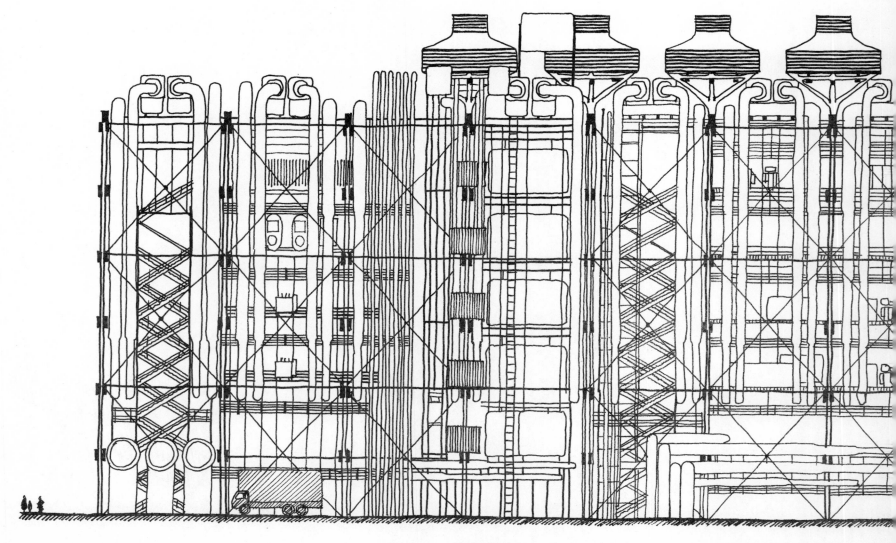

蓬皮杜艺术中心，巴黎，法国，理查德·罗杰斯和伦佐·皮亚诺，1977年

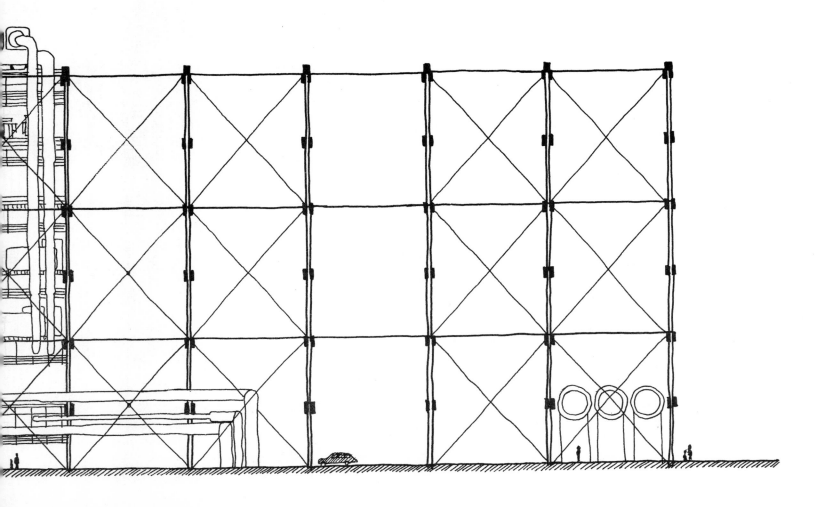

140

The form of many **modern museums** often express the buildings' content. The two examples here express two very different ideas: one is a work inspired by both sculpture and the surrounding former-shipyard area; the other is a powerful expression of the Jewish people's historical struggle.

许多当代博物馆的形式表达了建筑的内涵。此处的两个实例展示了两种不同的理念：一件作品是受到雕塑以及前造船厂环境的启迪；另一件作品强烈表达了犹太人民的历史斗争的情感。

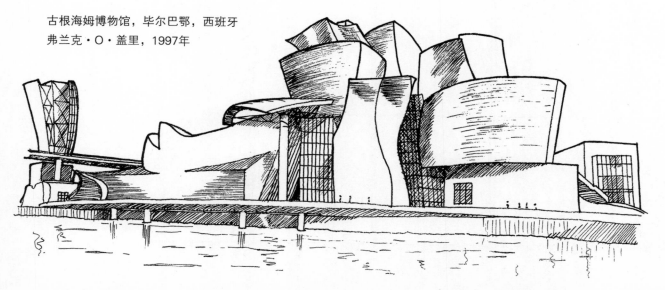

犹太人博物馆，柏林，德国
丹尼尔·里伯斯金，2001年

古根海姆博物馆，毕尔巴鄂，西班牙
弗兰克·O·盖里，1997年

设定一个博物馆的主题，绘制一些你认为最能体现主题精神的建筑造型……

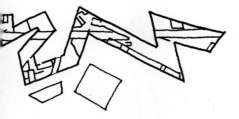

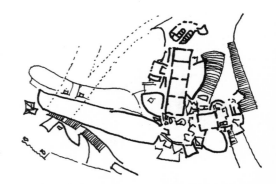

This mobile **Classroom of the Future** was designed by Gollifer Langston Architects in 2007. It can be transported by lorry from one location to another and provides workstations and audio/visual equipment.

2007年，戈利弗·兰斯顿建筑师事务所设计了移动式未来教室，它可以用卡车从一个地方运输到另一个地方，并且还可以提供工作站和视听设备。

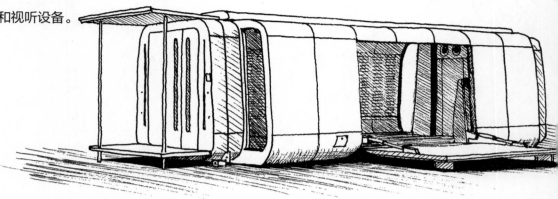

设计一个适合于标准卡车尺寸的移动式教育装置……

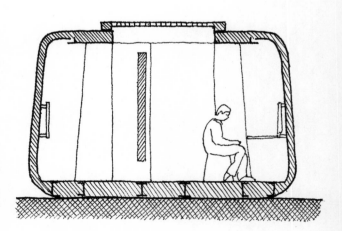

未来的教室，伦敦，英格兰，
戈利弗·兰斯顿建筑师事务所，2007年

Here is a selection of **designs for dogs** by different architects that will alter the way that the owners interact with their pets.

这是由不同建筑师设计的狗屋集选，这些设计将改变主人与宠物的互动方式。

贵妇犬的"高台"镜子，由康斯坦丁·格尔齐茨设计

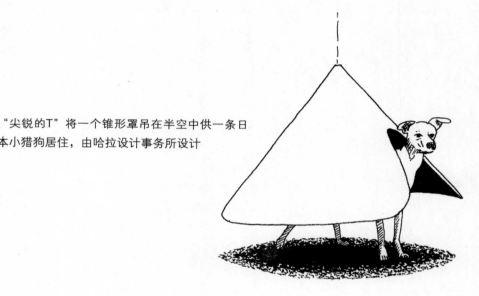

"尖锐的T"将一个锥形罩吊在半空中供一条日本小猎狗居住，由哈拉设计事务所设计

绘制一些家养动物的设计构思，如挂在散热器上的猫篮……

腊肠犬的滑梯，Bow-Wow工作室设计

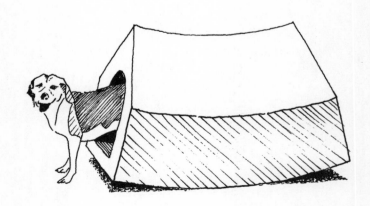

猎兔犬的摇摆狗舍，MVRDV设计

This prototype **circular house** used mass-produced aircraft technology and materials combined with an innovative rotating-roof ventilation system.

圆形住宅采用了批量生产的航空技术与材料，并创造性地运用了旋转屋顶的通风系统。

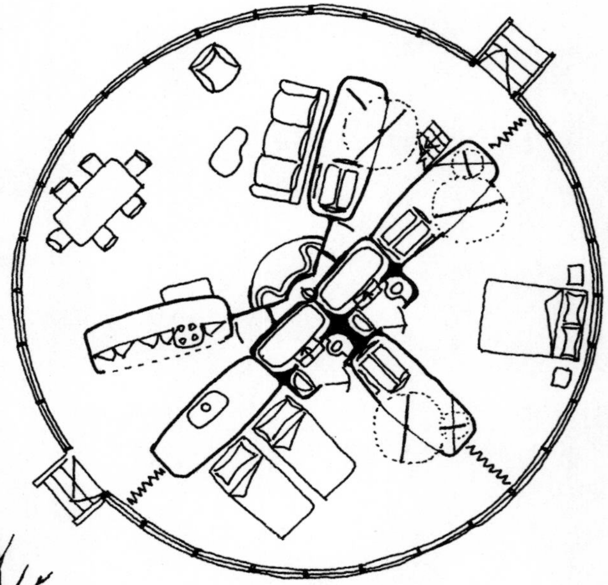

设计一个圆形住宅平面……

威奇塔屋，堪萨斯，美国，巴克敏斯特·富勒，1947年

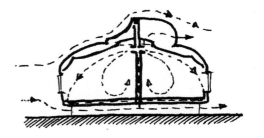

图示屋顶通风系统产生的空气流动

Incorporating water into the design of an environment can have both a dramatic and a calming effect.

景观中的亲水设计能够产生一种动人心弦而又宁静的效果。

绘制一个你熟悉的场所,花园、室内或公共场所,在其中设计一款新型的水体表现方式……

私人住宅,伦敦,英格兰
坦卡德·鲍克特,2007年

水之教堂,苫鹉,日本,安藤忠雄,1988年

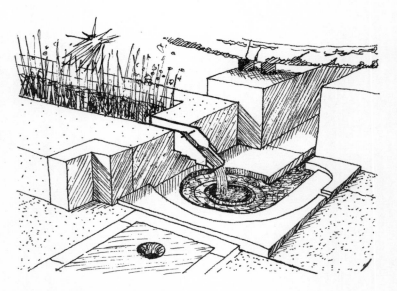

Querini Stampalia花园中的人造喷泉,
威尼斯,意大利,卡洛·斯卡帕,1963年

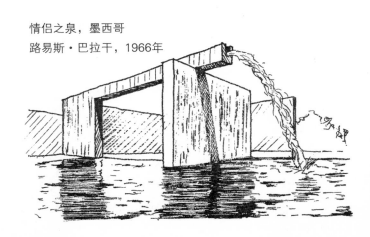

情侣之泉,墨西哥
路易斯·巴拉干,1966年

Frank Lloyd Wright described his work as **'organic' architecture,** 'where the whole is (to) the part, what the part is to the whole.' His Fallingwater house completely integrated the surrounding landscape as part of the environment of the house.

弗兰克·劳埃德·赖特将他的作品描述为"有机"建筑,即"整体属于局部,而局部又属于整体"。他的流水别墅完全与周围的环境景观融为一体,犹如环境中的一部分。

挑选一个特殊的场所,利用"有机建筑"原理,绘制一座住宅的设计草图,要求住宅与环境融为一体……

流水别墅,宾夕法尼亚,美国
弗兰克·劳埃德·赖特,1939年

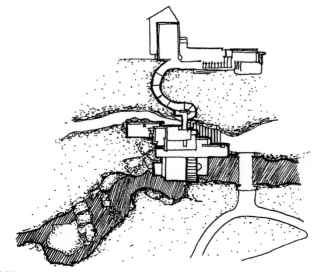

鸟瞰图

This is a **sectional drawing** of a hotel/spa that has been cut into the mountain over a thermal spring. The quartzite stone of the mountain is used to form the structure.

这是一张水疗酒店的剖面图，酒店跨越温泉并嵌入山体。山体的石英岩块石用于建筑结构承重。

用同样的山体斜坡，用剖面表示一个小的温泉浴室的设计，请考虑景观，光线和私密性……

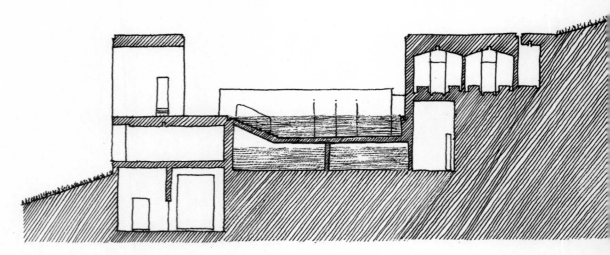

瓦尔斯温泉浴场，瑞士
彼得·卒姆托，1996年

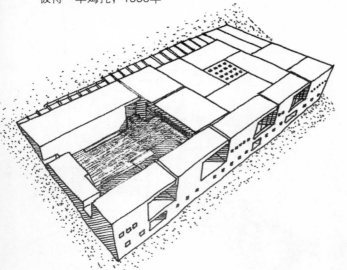

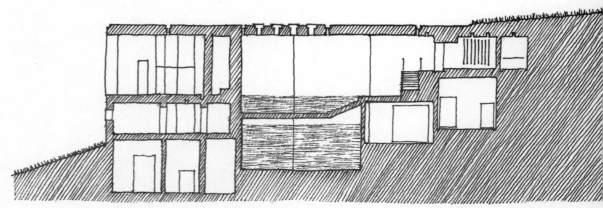

Breaking down **the threshold** between the street and gallery is the aim of this Storefront for Art and Architecture façade.

消解街道和画廊之间的入口界限是这家艺术与建筑铺面所追求的立面效果。

绘制一个新设计的方案，重点要考虑入口……

艺术与建筑铺面，纽约，美国
斯蒂文·霍尔和维托·阿科奇，1993年

Our major **cities of the future** are becoming taller, denser and a mixture of different styles and forms. This drawing of an imagined city is constructed using a variety of sampled projects and made-up forms and designs.

未来大城市逐渐变高、变密，成为各种风格和形式的混合体。这张设想的城市插图是用许多摘选的建筑与虚构的形式和设计共同勾勒绘制而成。

用你的远景构想完成这张城市全景图，完成后给城市命名……

158

You've completed the book and won the award for **Architect of the Year!**

你已经读完了这本书，并赢得了年度建筑师的奖品！

设计一种你很期望得到的奖品……

奥斯卡小金人；普利茨克建筑奖；英国电影学院奖；国际足联世界杯奖杯

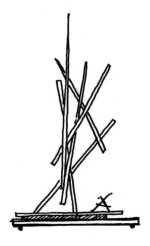

Credits & acknowledgements（致谢）

书中所有的插图都是作者专为本书而绘制。其中一些插图是依据建筑师和艺术家的原创性作品绘制而成，在此谨表谢意，名单如下。同时也要感谢版权所有者，如果有任何遗漏或错误请不吝赐教，出版社再版时将会给予更正。

在此，作者要感谢Jane Tankard一直以来给予的鼓励、启迪以及反馈意见，抱歉占用了他宝贵的时间。也要感谢Philip Cooper，Gaynor Sermc Laurence King出版社团队的帮助、支持和出版编辑。感谢Newman and Eastw 事务所的Matt Cox为本书装帧设计。

Sketch of Falkestrasse rooftop renovation by Coop Himmelb(l)au, 1988
Typeface by Theo van Doesburg
Mae West Lips Sofa, Salvador Dalí, 1937 © Salvador Dalí, Fundació Gala-Salvador Dalí, DACS, 2013
Prototype Bauhaus font by Herbert Bayer, Dessau Bauhaus, 1925 © DACS 2013
Plan of The Chapel of Note Dame du Haut by Le Corbusier, 1954 © ADAGP, Paris and DACS, London 2013
'The Sixth Order or The End of Architecture' by Leon Krier
Comparative study of corner houses by Rob Krier
'Architectural Machines' by Yakov Chernikhov, 1931
Suprematist Composition by Kazimir Malevich, 1916
Lenin Tribune by El Lissitzsky, 1920
'I am a Monument' by Robert Venturi and Denise Scott Brown, 1972
Sketch of proposed National Stadium, Japan by Zaha Hadid, 2019
Plan of The Farnsworth House by Ludwig Mies van der Rohe, 1951 © DACS 2013
Plan of the Berlin Building Exhibition by Ludwig Mies van der Rohe and Lilly Reich, 1931 © DACS 2013.
Sketches of The Einstein Tower by Erich Mendelsohn, 1921
Plan of Gut Garkau Farm by Hugo Häring, 1926 © DACS 2013
Plan of The Storefront for Art and Architecture, New York, by Stephen Holl and Vito Acconci, 1993 © ARS, NY and DACS, London 2013 (Acconci only)